THE
RELIABILITY
OF SENSE
PERCEPTION

OTHER WORKS OF WILLIAM P. ALSTON

Perceiving God:
The Epistemology of Religious Experience

Divine Nature and Human Language:
Essays in Philosophical Theology

Epistemic Justification:
Essays in the Theory of Knowledge

Philosophy of Language

Religious Belief and Philosophical Thought:
Readings in the Philosophy of Religion (editor)

Readings in Twentieth-Century Philosophy
(coeditor with George Nakhnikian)

The Problems of Philosophy: Introductory Readings
(coeditor with Richard B. Brandt)

THE RELIABILITY OF SENSE PERCEPTION

William P. Alston

Cornell University Press

Ithaca and London

First published 1993 by Cornell University Press.

International Standard Book Number 0-8014-2862-9 (cloth)
International Standard Book Number 0-8014-8101-5 (paper)
Library of Congress Catalog Card Number 92-54964

Printed in the United States of America

Librarians: Library of Congress cataloging information appears on the last page of the book.

⊗ The paper in this book meets the minimum requirements of the American National Standard for Information Sciences—Permanence of Paper for Printed Library Materials, ANSI Z39.48-1984.

For Norman Kretzmann

CONTENTS

PREFACE

This monograph is an outgrowth of a part of my book *Perceiving God* (Alston 1991a, copyright © 1991 by Cornell University). In arguing that putative experience of God can be a source of epistemic justification for certain kinds of beliefs about God, I found myself taking the position that it is impossible to give an effective noncircular demonstration of the reliability of any of our basic modes of belief formation. (Hence the fact that we cannot do this for religious experience is no special demerit of that source of belief.) Rather than try to argue this in detail for every way of forming beliefs, I took sense perception as a test case and devoted chapter 3 of the book to presenting a fairly detailed case for the impossibility of showing in a noncircular way that sense perception is a reliable source of belief about the external environment. It occurred to me, and I was encouraged in this thought by others, that many philosophers who would not open a book entitled *Perceiving God* would be interested in this line of argument. I decided to expand the treatment in chapter 3 of that book and publish it separately. Hence the present work. I have prefaced the central argument with some stage setting, and have appended to it a brief indication of where I think we should go from there—what attitude it is most reasonable to take toward the situation I have claimed to exhibit.

I have been thinking, off and on, for decades about the prospects for validating modes of belief formation, and my thoughts on the subject have been nourished by many more sources than I can recall. The most recent major contributions have come from Robert Audi,

Jonathan Bennett, Nelson Pike, Alvin Plantinga, and Philip Quinn. Special thanks go to Norman Kretzmann and to students in his seminar on the manuscript of *Perceiving God* at Cornell in Spring 1991, to the students of a somewhat earlier seminar of mine, to the participants in a week-long workshop I gave on the epistemology of religious experience at Syracuse University in the summer of 1985, to the faculty at Western Washington University with whom I discussed these matters in the spring of 1986, and to the participants in a National Endowment for the Humanities Summer Institute in Philosophy of Religion at Western Washington University in the summer of 1986.

WILLIAM P. ALSTON

Fayetteville, New York

THE
RELIABILITY
OF SENSE
PERCEPTION

Chapter 1

INTRODUCTION

i. The Problem

Why suppose that sense perception is, by and large, an accurate source of information about the physical environment? Why suppose that memory is an accurate source of information about the past? Why suppose that rational intuition (i.e., the awareness of self-evident propositions) is an accurate source of information about the realm of necessary truths? Why suppose that inductive reasoning is an accurate source of information about the regularities that obtain in nature? More generally, why suppose that any of the bases on which we regularly and unquestionably form beliefs are reliable, can be relied on to yield mostly true beliefs? The fact that we continually form beliefs in these ways with the utmost confidence shows that we are strongly disposed to believe that these familiar sources are reliable. But do we, or can we, have any solid reason for this assurance? Is it a groundless faith, an "animal faith" in George Santayana's striking phrase, however unquestioningly it is embraced? Or do we possess, or can we obtain, adequate reasons for supposing that this is the way things are?

This essay is directed to the sort of question just indicated. More specifically, I will concentrate on the question of the reliability of sense perception. I will examine the most important attempts to show that sense perception is a reliable source of beliefs about the environment. I shall argue that the attempts that are not disqualified on other grounds fall victim to a certain kind of circularity that is extremely difficult to avoid in enterprises of this sort, what I will term 'epistemic circularity.' I will conclude that until and unless more successful arguments are forthcoming we shall have to acknowledge that we are unable to provide the kind of backing for our confidence in perceptually generated beliefs that has traditionally been sought. I will also, though much more briefly, suggest that a like result would be obtained by analogous scrutinies of attempts to establish the reliability of our other basic sources of belief. Finally, I will, in a sketchy and programmatic way, confront the question of what attitude we should take toward the situation thus revealed, though a full consideration of this problem must await another occasion.

ii. Significance of the Problem

The central concern of this essay has by no means been neglected in the history of philosophy—and for good reason. First, the reliability of the ways in which we form beliefs is obviously an issue of the greatest moment, not only for philosophical reflection but also for the conduct of life. It is of the first importance to us, both theoretically and practically, that our beliefs be true. Insofar as our beliefs are false they will, at least in most cases and apart from unusual circumstances, provide poor guidance to our efforts to realize our goals.[1] And the theoretical interest in finding out the way things are is, obviously, better satisfied by having the truth! Since a source of belief is reliable if and only if it is so constituted that it would yield mostly true beliefs in situations of the sorts in which it is employed, the concern that our beliefs be true transfers to the concern that what produces our beliefs be reliable. Thus it is not surprising that the issue of the reliability of belief sources has attracted much philosophical attention, though it has not always been phrased in precisely these terms. From ancient times skepticism has been largely based on reasons for doubting the reliability of such belief sources as perception, memory, and reason-

---

[1] Recent attempts to deny, or cast doubt on, this apparent truism I find quite unconvincing. See, e.g., Stich 1990, chap. 5. We can certainly envisage situations in which, because of an unusual concatenation of circumstances, a false belief about a certain matter will lead us to act more adaptively than a true belief. Thus if Smith is planning to murder me during a meeting, a false belief as to the time and place of the meeting will be more profitable to me than getting it straight. But these are exceptions that stand over against a mass of cases in which we will do much better with true beliefs.

ing. Descartes launched modern philosophy by attempting to establish the reliability of rational intuition and a lesser reliability for sense perception. The so-called problem of the external world, which has bulked so large in the philosophy of the last few centuries, has largely been concerned with the question of whether our sense experiences provide a reliable basis for the beliefs about the external environment we base on those experiences.

But we should not suppose that philosophical interest in the reliability of doxastic practices stems solely from a concern with skepticism and with the obvious practical importance of reliability. There is also the question of how the reliability with which our beliefs are formed is related to "justified" belief and to knowledge. One prominent recent movement in epistemology (appropriately called 'reliabilism') takes the reliability of the way a belief was generated to be crucial to whether the belief is justified and, if it is true, whether it counts as knowledge. Prominent exponents of this point of view include David Armstrong, Alvin Goldman, Fred Dretske, Marshall Swain, and Robert Nozick. Armstrong, Dretske, and Nozick shy away from questions of justification and connect reliability directly with knowledge. Although the details of their formulations differ considerably, the idea they have in common could be roughly stated as follows: *"to know that p" is to have a reliably generated belief that p.*[2] Goldman and Swain, on the other hand, develop a notion of the justification of belief in terms of reliability. Again the details differ, but the rough idea is that being reliably generated is what confers justification on belief.[3] Justification is then taken to be at least a crucial part of what converts true belief into knowledge. It is obviously central to a reliabilist epistemology to determine, or to consider ways of determining, which modes of forming beliefs are reliable, or are sufficiently reliable to confer justification or to make a contribution to converting true belief into knowledge. But apart from this particular position in epistemology, the question of truth is almost universally recognized to be of overwhelming importance epistemologically. Even those, like Roderick Chisholm and Richard Foley, who deny that likelihood of truth or reliability of source is required for being justified in believing that p, still

[2]See Armstrong 1973, Dretske 1981, Nozick 1981.
[3]See Alvin Goldman 1979, 1986; Swain 1981.

recognize it as a constraint on the concept of epistemic justification that a belief's being justified is a good thing vis-à-vis the aim of believing what is true and avoiding believing what is false.[4] And if that is the central goal by reference to which candidates for the status of justification are to be measured, then the notion of reliable belief formation must also be of central importance to epistemic assessment. Whatever one's epistemological theory, one can hardly avoid recognizing the crucial importance of the reliability of belief formation.

iii. Ways of Belief Formation

I have been speaking indifferently of "sources" of belief and of "ways" (I might just as well have said "modes" or "procedures") of forming beliefs. The former locution is a natural one, but it doesn't mark out with sufficient precision what we need to assess in terms of reliability. A given source will, at best, reliably give rise to certain kinds of beliefs and not others. Nothing is a reliable source for beliefs generally, with the possible exception of an infallible and omniscient authority. For example, basing beliefs about fundamental physical laws or about the nature of God directly on sense experience would be highly unreliable. When sense perception is said to be a reliable source of belief, what is meant is that our usual ways of forming beliefs about the physical environment on the basis of sense experience (together, perhaps, with suitable background beliefs) are reliable ones. Thus talk of the reliability of a source is to be understood as shorthand for the reliability of a certain set of ways of forming beliefs that take something from that source as input. The formulation in terms of ways (modes, habits, mechanisms . . .) of belief formation is the more accurate one. In this book I shall, for stylistic reasons, sometimes speak of the reliability of a source. This is always to be construed as just indicated.

Where I speak of "ways" or "modes" of forming beliefs, other theorists speak of "processes", "mechanisms", "procedures", or "habits". The differences between these terminologies are of no importance to the present investigation. Whatever the lingo, the basic idea is that a given psyche at a given time has a number of relatively fixed

---

[4]See Chisholm 1977, 1989; Foley 1987, chap. 3.

dispositions to go from a certain input (beliefs or experiences or a combination thereof) to a belief output with a content that is a more or less determinate function of relevant characteristics of the input. We can think of such a disposition, habit, mechanism . . . as the "realization" of a function that yields a belief content related in a certain way to relevant features of whatever is taken as input.[5] Where the input is at least partly made up of beliefs, the relevant characteristics will include the propositional content of those beliefs, or certain features thereof. Experiential inputs will display various features, some of which will be picked up in accordance with the embodied function as at least partly determinative of the content of the doxastic output. Thus if the input is the belief pair [John is a philosopher/ Philosophers are likely to be unathletic], a familiar function will yield a belief with the content [John is likely to be unathletic]. If the input is a sense experience that is an apparent visual display of a red vase-shaped object, then, assuming that there is no input or background indicating anything untoward in the situation, a familiar function will yield a belief of the form [That's a red vase] or [A red vase is in front of me]. If the input is a visual display that I would describe as "looking like a toy village" plus the belief that I am 35,000 feet above the object in question, again a familiar function would yield as output the belief that what I am looking at is a real village.

Doxastic (belief-forming) functions differ in the width of the input and output types involved. The input type could be something as narrow as a certain determinate configuration of specific sensory qualia and the output type something as narrow as a belief to the effect that the object in the center of the visual field is Susie Jones. Again on the narrow side, the input type might be a belief of the form 'X is an American citizen' and the output type a belief of the form 'X is not an Israeli citizen'. Or the function could be wider in scope. In thinking about perceptual belief formation we might specify a very wide function that takes inputs of the type *an experience of the sort S would be inclined to take as a case of X's appearing ϕ to S, and yields outputs of the correlated type a belief of the form "X is ϕ".* Among inferential mechanisms we have functions corresponding to each form (invalid as well

---

[5] I am not assuming that these functions are completely determinate. There may be a certain degree of free play in the details of the output that is yielded from a certain input.

as valid, I am afraid) of inference. Thus we can have a modus ponens function that takes as input a belief of the form ' If p then q, and p', and yields an output of the correlated form 'q'. It is a difficult question in cognitive psychology as to just what belief-forming mechanisms there are in the psyche and how wide or narrow they are. But for epistemic purposes we can think of individual mechanisms as quite narrow in scope. If one can reliably go from a certain specific pattern of sensory qualia to the attribution of a certain property to the current object of perceptual attention, then that attribution will be justified whether the most adequate psychology would ascribe that particular transition to a separate mechanism or construe it as a particular application of a more general mechanism.

We may distinguish between "generational" and "transformational" mechanisms. Generational mechanisms produce beliefs from nondoxastic inputs; they can introduce radically new content into the belief system. Transformational mechanisms yield belief outputs from belief inputs. I have just noted that a mechanism may be mixed, working on both doxastic and nondoxastic inputs. Since the presence of a nondoxastic input provides the chance for new content in the belief system, even where there are also doxastic components of the input, it will be useful to classify these mixed cases as generational.

I should point out that since I am not concerned in this essay to put forward the reliability with which a belief is formed as a necessary condition of the belief's being *justified* or counting as *knowledge*, I am not faced with certain well-advertised problems that confront such views. If I am to determine whether a given belief is reliably formed, I must determine what general mechanism, habit, or disposition is such that its activation was responsible for that particular belief. For reliability or the reverse can be predicated only of something that is capable of repeated utilizations, so that there is room for the question of what proportions of its outputs are or would be true. And so I must identify which of a number of candidates is *the* general mechanism that was responsible for this particular product. And there are well known difficulties in carrying this through.[6] I am confident that these difficulties can be overcome, but I need not go into that here. For I am not concerned here with attributing *reliability of formation* to particular

[6]See, e.g., Feldman 1985.

beliefs. Rather, my question concerns doxastic mechanisms that are susceptible of repeated utilizations, or rather large families of such mechanisms. From the outset we are working at the level of general procedures that can produce a run of cases and hence can be said themselves to be reliable or the reverse. We don't have the problem of picking, from various general types, the one of which a particular belief formation is best regarded as a token.

To be sure, we are properly concerned not only about the re-liability of belief-forming mechanisms, but also the reliability of belief-preserving mechanisms. Much of the advantage of forming true beliefs would be dissipated if we did not, by and large, retain those true beliefs for use as needed. However, the question of the reliability of belief-*forming* processes is obviously the more basic one. If beliefs are not formed reliably, we need not bother about the accuracy of their preservation. This essay will be restricted to belief formation, even though belief preservation deserves full attention in a complete epistemology.

iv. Doxastic Practices

As the title indicates, the main protagonist in this essay will be our customary ways of forming beliefs about the external environment on the basis of sense perception. Let's call the assemblage of those ways 'sense perceptual practice' ('SP' for short).[7] I use this terminology against the background of views developed elsewhere,[8] to the effect that we engage in a variety of doxastic (belief-forming) practices that are ineluctably rooted in our lives.[9] A prominent member of this group is the practice of going from sense experience (together, some-times, with relevant background beliefs) to beliefs about things, events, and states of affairs in the immediate physical and social environment. Thus I have an experience that I would be disposed to describe as something in front of me looking like a birch tree, and on

[7]Note that contrary to what the terminology would naturally suggest, 'SP' is used, not for sense perception itself, but for the activity of forming beliefs (perceptual beliefs) about the physical environment on the basis of sense perception.

[8]Alston 1989c, 1991a, chap. 4.

[9]The use of the term 'practice' is not intended to carry the implication that the activities constituent of a practice are under voluntary control. The term ranges over any regular way of carrying out functions, especially psychological and behavioral functions.

that basis I believe that there is a birch tree before me. I see something that displays the characteristic look of your house, and on that basis, together with the realization that I am on the block on which you live, I form the belief that that is your house. I use the acronym 'SP' to range over all our customary ways of going from sense experience (together with background beliefs where that is involved) to beliefs about the physical and social environment. (Call such beliefs, when based on sense experience, 'perceptual beliefs'.) It is with the reliability of SP that we will be centrally concerned in this essay. Among other familiar doxastic practices are the formation of beliefs on the basis of memory, introspection, and rational intuition, and on the basis of deductive and non deductive reasoning.

Clearly, what I am calling a 'doxastic practice' is not a single belief-forming disposition, but some family, grouping, or system of individual dispositions, bound together in some important way. What binds the components together in the practice is some marked similarity in input, output, and/or function. Dispositions in the SP grouping obviously have in common the fact that the input is, at least in part, sensory experience. There are also important similarities in functions, though the presence of background beliefs of many different kinds and modes of relevance makes a simple formulation of the commonality impossible. A strong thread is the connection between 'looks P' and 'is P'. Leaving background beliefs out of the picture, we can say that a common type of function here is one that goes from 'looks P' to 'is P' (e.g., from something's looking red to the belief that it is red), in the absence of any reason to suspect that output.

v. Reliability

Now for a few words about the concept of reliability. First, to call a belief-forming mechanism 'reliable' is to judge that it will or would yield mostly true beliefs. But over what range of employments? Those in which it has been employed up to now? That would be to identify reliability with a favorable track record, but doing so can't be right. An unreliable procedure might have chanced to work well on the few occasions on which it was actually employed. Anyone can get lucky! If there have been only five crystal-ball readings all of which just happened to be correct, that wouldn't make reading a crystal ball a reliable way of forming beliefs; it might still have a poor record over the long haul. Indeed, we can't identify reliability with a favorable record over all past, present and future employments. A practice or instrument that is never employed might be quite reliable in that it *would* yield mostly true beliefs in the long run. Thus to call something reliable is to speak about the kind of record it *would* pile up over a suitable number and variety of employments. An actual track record is crucial evidence for judgments of reliability just to the extent that it is a good indication of that. But what makes a run of cases suitable? Briefly, the class of cases must be sufficiently varied to rule out the possibility that the results are due to factors other than the character of the procedure. Moreover, they must be the sorts of cases we in fact typically encounter. The fact that sensory experience would not be a reliable source of belief in unusually deceptive environments, or in cases of direct brain stimulation, does not show that standard perceptual belief formation is not reliable. So, to put it in a nutshell, a doxastic mechanism is reliable provided *it would yield mostly true beliefs in a sufficiently large and varied run of employments in situations of the sorts we typically encounter.* This is less than perfectly precise (for example, what does it take for a type of situation to be typical for us?), but it has just the kind of looseness we need for the purpose.

Second, what degree of reliability is in question? I have said that I am concerned with whether one or another way of forming beliefs "is reliable", but reliability is obviously a matter of degree; one instrument, method, or procedure may be more or less reliable than another. I have said that a mechanism is reliable (to put it shortly) if it would yield mostly true beliefs. But how much is "most"? I won't try to give a precise answer; I don't think there is any basis for doing so. What we are typically interested in—whether for justification, knowledge, or whatever—is whether the mechanism would yield a "high proportion" of truths.[10] But just how high a proportion we require may differ for different mechanisms, depending on, among other things, the degree of reliability it is realistic to expect. For example, the vision of objects directly in front of one has the capacity for a

[10]For some discussion of this issue see Alvin Goldman 1986, sec. 5.5. Of course, anything deserving of the title of 'most' or 'high proportion' would be considerably more than half, but that leaves ample room for variation.

greater degree of reliability than does the memory of remote events in one's early years. But we shouldn't deny that beliefs generated in the less reliable ways can thereby be justified or constitute knowledge. Perhaps we should distinguish degrees of justification in correlation with degrees of reliability, but I won't get into that.

Third, we should not be too fussy about exact truth in asking about reliability. Suppose that, as many philosophers have thought, typical perceptual beliefs are not exactly true as they stand. For one thing, they indefensibly represent such qualia as colors as attaching to the objects themselves. For another, they depict objects as unbrokenly continuous that are actually constituted by many tiny particles moving about in empty space. To be sure, this position is highly controversial. Many thinkers deny that ordinary perceptual beliefs are committed to any such construals; and others take it that perceptual beliefs as just characterized are strictly true. My aim here, however, is not to settle these issues but to use the view that ordinary perceptual beliefs are not strictly true in order to illustrate a point. That point is that even if perceptual beliefs misrepresent the environment in ways like these, they can still be highly useful guides to that environment. The real facts of the matter can be in a systematic correspondence with the colors and the unbroken surfaces that the perceiver mistakenly attributes to the things in themselves; so that what we believe on the basis of perception can be close enough to the truth for practical purposes. Thus, as I construe reliability, sense perceptual doxastic mechanisms could still be judged reliable if their outputs are usually close enough to the truth, even if they do not strictly hit the mark.

It is important to distinguish our question as to how, if at all, the reliability of SP can be established from others with which it might be confused; for a successful answer to one of these other questions will not necessarily answer ours. I noted earlier that questions about the reliability of sense perception and other familiar sources of belief have figured heavily in skepticism and in controversies to which it gives rise. It might, therefore, be thought that an attempt to establish the reliability of SP is essentially and necessarily an attempt to "answer skepticism" or to "resolve skeptical doubts" about SP. But though this is a context within which, and a purpose for which, people have often sought to validate the reliability of SP, it is not necessarily so. In asking whether SP is reliable, in seeking to show that it is, or, as we are doing, seeking to determine whether it is possible to show that, we are not necessarily presupposing that there is any real *doubt* about the reliability of SP. We are seeking only to determine what, if anything, can be done to *show* that it is reliable. Hence dismissal of skeptical doubts has no crucial bearing on this enterprise. For example, one response to skepticism about perception is that it is exaggerated, unreal, artificial, and not to be taken seriously. But even if that is a valid basis for dismissing doubts about the reliability of SP, it does nothing to answer the question as to whether it is possible to *show* that SP is reliable.

Second, many of the discussions we will be considering involve attempts to demonstrate the existence of the "external world" or the "physical world", starting merely from premises concerning one's own conscious experience (and perhaps a priori truths). But establishing the existence of a world, even a physical world, beyond our experience does not suffice to show that SP is reliable. Solipsism is not the only alternative to the reliability of SP. It could be that there is a physical world beyond our experience but that our sense perception does not serve as an accurate guide to it. Finally, some Wittgensteinian arguments we will be considering are concerned to show that SP could not *always* be mistaken. But even if successful in that, those arguments fall short of showing that SP is *reliable*. Reliability takes more than the production of the occasional truth.

Chapter 2

TRACK RECORD AND OTHER SIMPLE EMPIRICAL ARGUMENTS FOR RELIABILITY

i. A Track Record Argument for the Reliability of Sense Perceptual Practice (SP)

Against this background we are ready to tackle our central question, whether it is possible to establish the reliability of SP. Let's begin by considering generally what it takes to determine whether a way of forming beliefs is reliable. The most obvious approach is inductive. Check a suitable sample of outputs for truth and take the proportion of truths in that sample as an estimate of the reliability of that mode of belief formation. Though, as just pointed out, we cannot *identify* reliability with a favorable track record, such a record in a suitably large and varied spread of cases is the best and most direct evidence for reliability. Furthermore, it would seem that any other evidence would be parasitic on this. Suppose, for example, that a perceptual doxastic mechanism that takes into account background information of relevant sorts—where the subject is, how situated relative to the object, etc.—is more reliable than one that does not. How would we know this to be the case? Presumably, by determining that mechanisms of the former sort have a better track record than mechanisms of the latter sort (when tested in comparable situations), or else by correlating this difference with other differences that indicate differential reliability, in which case the assumption of this latter differential reliability must be validated by a differential track record, or . . . Thus, although there is no limit to the variety of indicators of degree of reliability, the most direct indication—favorable track record—is fundamental to all the others. Hence it will be appropriate to focus on it in these preliminary considerations.

If we try to assess the reliability of SP in this way, however, we immediately run into a roadblock. How do we determine whether the perceptual beliefs in our sample are true? If we do so by taking another look, listen, or whatever, we would seem to be presupposing the reliability of SP in order to compile evidence for that reliability and so to fall into circularity. Moreover, more complex ways of determining the truth value of perceptual beliefs will run into the same problem, though less directly. The basic point here is that any way we have of determining perceivable facts in the physical world will depend, sooner or later, on what we learn from sense perception. Even when what we appeal to in showing that X is or is not P (that this wine is a Chambolle Musigny) is not itself a perceptual report (that the wine in this bottle came from a cask used to store wine made from grapes grown in the commune of Chambolle Musigny), still, in showing that that is the case, we must at some point rely on perceptual reports. We may be relying on various records that detail the historical origin of the wine in this bottle. Not only must we use our eyes to read the reports, but the reports themselves, if valid, were based on observations of the grapes and the wine at various stages of their history.

I don't know of any simple knockdown argument for this thesis, that any cognitive access we have to the physical world ultimately rests on sense perception. All I can do to support it is the following. First, I can go through a number of cases like the one just mentioned and exhibit the plausibility, in each case, of the claim that SP must be relied on at one or more points. Second, we can survey the resources we have for determining particular perceivable matters of fact, and verify that they all involve reliance on SP at some stage. Thus, in addition to a direct appeal to perception of the putative fact in question, there are the following. (1) Memory of past perceptions, which is, of course, only as trustworthy as the perceptions remembered. (2) Inferences from particular past facts plus general regularities, as in the Chambolle Musigny case. Here it seems clear that our evidence for both the past facts and the generalizations depends on sensory observation and hence on SP at some point. (3) The use of instruments. Here we must rely on SP both for reading the instrument and for our evidence for its reliability. (4) The use of reliable indications, as when we know from a vapor trail in the sky that a jet plane has flown by recently. Here we rely on SP both for the knowledge of the vapor trail and for the generalization connecting jet flights and vapor trails. (5) More theoretical inferences, as when the presence of another planet was inferred from deviations in the expected orbit of observed planets. Here we rely on SP both for the observation of the known planets and for evidence for the theory used in the derivation. And so it goes. Finally, there is the general recognition that we have no a priori access to particular perceivable facts. Even if it were the case, and this is dubious, that we can establish a priori certain very general principles about the physical world, like 'Every event has a cause' or 'Energy is conserved through every physical transaction', it still remains the case that I cannot by mere thinking determine whether your car is in the garage now or that you are now wearing a red dress.

Let's assume, then, that it is impossible to determine that a particular perceptual belief is true or false without making use at some point of what we take ourselves to have learned from SP. It follows that we cannot use an inductive track record argument for the reliability of SP without presupposing that reliability, and so falling into circularity. Let's use the term 'basic' for any doxastic practice of which this is true, that is, any practice for the reliability of which any otherwise effective track record argument would be circular. Such a practice is properly called 'basic', since its epistemic claims cannot be validated solely by the use of other practices. Thus, as I shall indicate more fully in the sequel, it has a claim to be at the "basis" of our cognitive endeavors. It represents a way of forming beliefs that cannot be judged, at least in any simple fashion, on the basis of the output of other, more fundamental practices.

To be sure, I began by saying that our central issue was whether it is possible to determine the reliability of SP, that is, to determine *whether or not* it is reliable. And in exploring the possibility of a track record argument, all I have shown is that we fall into circularity if we use it to establish reliability. But what about its use to establish unreliability? There we could hardly make a charge of circularity stick. If the conclusion is that SP is unreliable, I could not claim that my use of perceptual premises reflects my practical assumption that SP is *unreliable*. Quite the contrary. If such an argument also relies on SP as a source of information, the charge would be different though equally disqualifying, namely, that the argument presupposes the contradictory of the conclusion. But we are not in a position to claim that any otherwise effective argument for the unreliability of SP relies on SP in this way. For it would be an effective argument for the unreliability of a doxastic practice to exhibit enough inconsistencies between its outputs. We wouldn't have to determine what's what with respect to its subject matter in order to conclude on those grounds that the practice is unreliable. In fact, though perceptual reports do sometimes contradict each other, the incidence of that seems clearly not to be of such a magnitude as to justify taking SP to be significantly unreliable, though it certainly does show that SP is not perfectly reliable. In any event, there is little inclination among contemporary philosophers to take seriously the idea that SP is radically unreliable. Hence, in this discussion I shall ignore that side of the matter and confine my attention to attempts to establish the reliability of SP.

ii. Epistemic Circularity

This is a good place to take a harder look at the kind of circularity that is involved in the track record argument. It is not the most direct kind of logical circularity. We are not using the proposition that sense perception is reliable as one of our premises. Nevertheless, we are assuming the reliability of sense perception in using it, or some source(s) dependent on it, to generate our premises. If one were to challenge our premises and continue the challenge long enough, we would eventually be driven to appeal to the reliability of sense perception in defending our right to those premises. And if I were to ask myself why I should accept the premises, I would, if I pushed the reflection far enough, have to make the claim that sense perception is reliable. For if I weren't prepared to make that claim on reflection, why would I, as a rational subject, countenance perceptual beliefs? Since this kind of circularity involves a commitment to the conclusion as a presupposition of our supposing ourselves to be *justified* in holding the premises, we can properly term it 'epistemic circularity'.

Epistemically circular arguments would seem to be of no force. If we have to assume the reliability of SP in order to suppose ourselves entitled to the premises, how can an argument from those premises, however impeccable its logical credentials, provide support for that proposition? But surprisingly enough, as I argue in "Epistemic Circularity,"[1] that does not prevent our using such arguments to *show* that sense perception is reliable or to *justify* that thesis. Nor, pari passu, does it prevent us from being justified in believing sense perception to be reliable by virtue of basing that belief on the premises of a simple track record argument. At least this will be the case if there are no "higher level" requirements for being justified in believing that p, such as being justified in supposing the practice that yields the belief to be a reliable one, or being justified in supposing the ground on which the belief is based to be an adequate one. And, though I cannot defend the position here, I do not think of justification as subject to any such requirements. On my view, a belief is justified if and only if it is based on an adequate ground; that is, it is necessary only that the ground *be* adequate, not that the subject know or justifiably believe this, much less that the subject know or justifiably believe that all requirements for justification are satisfied.[2] But then I can be justified in accepting the outputs of a certain doxastic practice without being justified in believing that the practice is reliable. I need not already be justified in supposing SP to be reliable in order to be justified in various perceptual beliefs. SP must *be* reliable if I am to be justified in holding perceptual beliefs, but I don't have to be justified in supposing this to be the case. But then *if SP is reliable,* I can use various (justified) perceptual beliefs to show that SP is reliable, for I need not already be justified in holding the conclusion in order to be justified in holding the premises. The argument would still be *epistemically circular,* for I am still assuming *in practice* the reliability of SP in forming normal perceptual beliefs. Nevertheless, I don't have to be

[1]Alston 1989b.

[2]For a defense of this position see "An Internalist Externalism" in Alston 1989b and 1991b. One reason, and a conclusive one, for avoiding such higher level requirements is that they imply that one is required to have an infinite hierarchy of justified beliefs. For if to be justified in the belief that p one has to be justified in believing that *the ground of the former belief is an adequate one* (call the italicized proposition 'q'), then a parallel condition will be put on the belief that q; to be justified in holding it one must be justified in believing that *the ground of the belief that q is an adequate one.* And to be justified in that belief, one will have to have a still higher level justified belief that the ground of that belief is adequate. And so on ad infinitum. Since it seems clear that no human being can possess an infinite hierarchy of beliefs-justified or not-to impose any such requirement will imply that no human being justifiably believes anything, and our subject matter will have disappeared.

justified in making that assumption, in order to be justified in the perceptual beliefs that give me my premises. Hence the epistemic circularity does not prevent justification from being transmitted from the premises to a conclusion that would have been unjustified except for this argument. That applies even to a simple track record argument.[3]

But even if I am right about this and it is possible to establish the reliability of sense perception and other basic sources of belief by simple track record arguments, these arguments still do not satisfy the usual aspirations of those seeking to determine whether a basic doxastic practice like SP is reliable. The reason is this. What I pointed out in the previous paragraph is that *if sense perception is reliable,* a track record argument will suffice to show that it is. Epistemic circularity does not in and of itself disqualify the argument. But even granting that point, the argument will not do its job unless we *are* justified in accepting its premises; and that is the case only if sense perception is in fact reliable. This is to offer a stone instead of bread. We can say the same of any belief-forming practice whatever, no matter how disreputable. We can just as well say of crystal ball gazing that if it *is* reliable, we can use a track record argument to show that it is reliable. But when we ask whether one or another source of belief is reliable, we are interested in *discriminating* those that can reasonably be trusted from those that cannot. Hence merely showing that *if* a given source is reliable it can be shown by its record to be reliable, does nothing to indicate that the source belongs with the sheep rather than with the goats. I have removed an allegedly crippling disability, but I have not given the argument a clean bill of health. Hence I shall disqualify epistemically circular arguments on the grounds that they do not serve to discriminate between reliable and unreliable doxastic practices.

iii. A Piecemeal Approach

Thus, a simple track record argument cannot give us what we are looking for when we attempt to show that SP is reliable, just because it is infected with epistemic circularity. Faced with this result, one might suggest that the trouble is that we have chosen too large and

[3]See "Epistemic Circularity" (Alston 1989b) for much more detail on this argument.

unwieldy a unit for our investigation. Why not consider the question of whether some subset of perceptual belief-forming mechanisms is reliable? In that case, there is no reason to think that we have to presuppose, even in practice, the reliability of the ways being tested in order to pile up evidence for their reliability. To explore this, let's start at the other end of the spectrum and consider a very specific belief-forming mechanism, for example, one that takes something's looking like a peach as input and yields a belief [That's a peach] as output. To take a track record approach to establishing reliability here, we consider a sufficient number of sufficiently varied cases in which this input-output function was operative, cases that are all "normal" in the sense of being in circumstances of the sort we frequently encounter (no laser images, no Cartesian demon). We then determine in what proportion of those cases the belief engendered was true, that is, in what proportion of those cases the object in question was a peach. And how do we do that? Obviously, we have to utilize some other way of spotting peaches. We can't use this mechanism, whether in the original subject or others, without begging the question. And we do have other ways of doing this. We can smell, taste, and feel the object. We can determine whether it was picked from a peach tree. We could run biochemical and even microbiological tests. We can make ice cream or pie from it and see what that tastes like. And so on. In none of these cases would we be relying on the move from *It looks like a peach* to *It is a peach*. To be sure, in these cases we are relying heavily on other perceptual mechanisms. Even in the sophisticated biochemical and microbiological tests, the tester has to rely on what she sees, hears, or otherwise perceives at some point in carrying out the test. But that in no way casts any doubt on our capacity to determine the truth of the outputs of *this* mechanism without circularity. For at no point do we appeal to beliefs generated by that mechanism itself.

Thus it is clear that where we are testing a very specific source of belief for reliability, there is no difficulty in principle in carrying out a track record test without falling victim to circularity. And the source does not have to be nearly this restrictive to achieve this result. The same point can be made about testing visual color or shape attributions generally, or visual identification of objects generally, or even visual belief formation generally (and so for the other senses). To be sure, the more extensive the mechanism (or group of mechanisms) the more difficult it is to find an independent check on truth. For visual shape attributions (just on the basis of how the object looks) we have a variety of other checks, including how it feels as well as measurements of various sorts. But if the group is as large as visual beliefs generally—even restricting ourselves to "primary" visual beliefs, beliefs about objects formed just on the basis of how they look—things are more difficult. To carry through a noncircular track record argument here we would have to determine in each case whether X is, for example, a birch tree without relying at any point on how something looks. Testing procedures that rely on appearances only to the other senses could be devised, I suppose, but doing so might require considerable ingenuity.

iv. Back to SP as a Whole

Thus, whenever it is a question of the reliability of some mode (or family of modes) of perceptual belief formation that is more restricted than the whole of that territory, we can find ways of checking its outputs for truth value that are independent of any reliance on the mode(s) under scrutiny. But on further reflection it can hardly fail to strike us that we will run into circularity if we persist long enough. To oversimplify, let's say we check the accuracy of visual beliefs by using audition and touch in the perceptual part of those tests.[4] Now what happens when we test the reliability of auditorily and tactilely formed beliefs? Perhaps we can get by with the other two senses (though I doubt it), but even if we could, when we come to test them we will perforce have to rely on one or more of the modalities already checked. So, to put it schematically, vision is validated by audition and touch. Touch is validated by smell and taste. Smell and taste are validated by vision and audition. And there we have our circle. The details of the procedure will change the relative position of the modalities on the perimeter of the circle, but they cannot change the general character of the proceedings. The basic point is that we cannot do anything toward determining the truth value of singular statements about

---

[4]As this example indicates, I am not supposing, and it is not the case, that in determining the truth value of a perceptual belief we rely solely on perception. We also will make use of reasoning of various sorts and we will rely on memory. Here I am concerned only with the point that perceptual beliefs constitute an essential part of what we are using and trusting.

perceivable objects without relying on sense perception for information somewhere in the proceedings. More precisely, we cannot do so without relying on beliefs the justification of which rests, proximately or remotely, on sense perception. If I already know that all the trees on my property are birches, then that confirms a perceptual report that an object in a certain place on my property is a birch tree; I don't have to take another look myself. But, of course, I have that antecedent knowledge only because I have seen the trees many times and recognized them as birches, or because someone who was in a position to know told me that all those trees are birches and he knew that because he visually recognized them as birches; or . . . Thus it seems clear that any basis we have for supposing a perceptual belief to be true or false will involve, somewhere along the line, a reliance on the accuracy of perceptual belief formation. And that being the case, though we can support the reliability of certain stretches of the territory by relying on other stretches, we will, since the variety of such stretches is finite, eventually be involved in a circle in seeking to establish the reliability of any part of the territory.

I had better seize this opportunity to make explicit a complication that affects many of the points I will be making about the epistemic circularity of various arguments for the reliability of SP. From the standpoint of coherence theory this circularity is not a worry, provided the circle is large enough. Coherence theory holds that individual beliefs gain positive epistemic status (justification, rationality, being "established", or whatever) by virtue of being involved in a total system of beliefs that is coherent. And to say that a system is coherent is to say, roughly, that there are strong bonds of mutual support between the constituents. There is no unidirectional support that passes from "foundations" that need no support from other beliefs to the rest of the system, the "superstructure". Rather, every component belief and every subsystem is involved in both supporting and being supported by others. Reciprocal support is the rule rather than the exception. From this standpoint there is nothing disturbing about the circle involved in using perceptual beliefs to support the principle that sense perception is reliable. Particular perceptual beliefs, on the one hand, and the belief in the reliability of sense perception, on the other, support each other, thereby increasing the coherence of the system that contains both. However, for purposes of this essay I am setting aside coherence theories without a hearing. I take it that the live possibility of a multiplicity, perhaps an indefinite multiplicity, of incompatible but equally coherent systems of equal comprehensiveness is sufficient to show that internal coherence cannot be the whole story of what gives beliefs a positive epistemic status. And therefore I will continue to regard circularity as disqualifying.

The above discussion amounts to one way of motivating a concern with the reliability of sense perception generally. For if the attempt to justify a particular group of perceptual doxastic mechanisms by relying on others to do so inevitably leads to a circle of self-support within the sphere of perceptual belief formation, it looks as if such efforts presuppose the general reliability of SP. If perceptual belief is generally reliable, then it is in order to check on any suspicious part thereof by relying on other parts about which suspicions have not been aroused. But if we are not already entitled to take sense perception to be mostly reliable, the fact that some of its outputs confirm others would seem to be of little epistemological significance. For all we would have any reason to suppose, it might be one of those vast coherent systems of fancy that are regularly thrown up as an objection to the idea that the coherence of a system is a sufficient indication of its truth. Hence we are driven to raise the question of reliability about sense perception generally. And once that general question is raised it is obvious that we cannot use perceptual beliefs (perceptually derived putative information or knowledge) as any part of our case for a positive answer to the question. That would be a circularity more obvious than the one we saw in the piecemeal approach to the epistemic assessment of sense perception. Hence we are ineluctably led to the basic issue of this essay: is it possible to establish the reliability of sense perception in a noncircular fashion?

Nor does the interest of this larger question wholly stem from the background I have just sketched. Sense perception as a whole is a natural unit about which to raise the question of reliability. Though I suppose that it is conceivable in the abstract that, say, vision gives us generally accurate information but audition or touch does not, and though there obviously is a difference in the reliability of different perceptual mechanisms (up close vision is more reliable than distant vision), still it seems overwhelmingly plausible to suppose that sense perception stands or falls as a whole at the bar of critical scrutiny. It is difficult to take seriously the idea that one of our sense modalities might be *radically* different from others in the reliability of the information it provides, or that one of the senses might vary *widely* in its reliability as a source of the beliefs it regularly and confidently engenders (for example, that vision might be reliable for the identification of trees but not of animals). Hence it is quite understandable that the question of the reliability of sense perception *überhaupt* has been a traditional preoccupation of philosophers.

Another reason for this preoccupation is that the major reasons for doubting the reliability of sense perception have always seemed to apply across the board, not to some senses rather than others, or to some particular kinds of beliefs. This is true of arguments from perceptual errors (if one can be mistaken some of the time one can be mistaken all of the time), from dreams or deceiving demons (we can't distinguish the case where we are only dreaming that the Louvre is in front of us, or where a demon is deceiving us into supposing we see the Louvre, from the situation in which we are veridically seeing the Louvre), and from underdetermination by sensory evidence (our sense experiences are quite compatible with things not being as they seem to indicate; they do not provide conclusive grounds for the beliefs we typically base on them). Thus insofar as concern with the reliability of the senses stems from skeptical doubts like these, it is understandable that the question should be addressed to the senses generally.

v. Pragmatic Arguments: Validation by Fruits

Where are we to turn next? A popular line of thought is that SP proves itself by its fruits, particularly by the way in which it puts us in a position to predict and thereby, to some extent, to control the course of events. It provides us with data on the basis of which we establish lawlike generalizations, which we can then use as the basis for prediction and control. By taking sense experience as a guide to what is around us we can learn in each of a number of instances that milk sours more slowly when cold than when warm. This puts us into a position to predict that a refrigerated bottle of milk will last longer than an unrefrigerated one, and we can use this knowledge to control the condition of our milk. This is the humblest of examples, and the predictive power is greatly increased in scope and precision as we move further into the higher reaches of theory; but the general point is the same. SP proves itself by what it puts us in a position to do. If it weren't usually giving us the straight story about what is happening around us, how could we have so much success in predicting the further course of events when we use its deliverances to build up systems of general beliefs that we use to make those predictions?

That sounds right. But how do we know that we are often successful in prediction? By induction from particular cases of success, obviously. But how do we know that we are successful in particular cases? By using our senses to determine whether what was predicted actually occurred, or by having recourse to some other method that sooner or later relies on the deliverances of SP. It is not as if an angel tells us that the prediction is borne out, or as if rational intuition does the job. But then we are back in epistemic circularity. We can mount this argument for the reliability of SP only by using SP to get some of our crucial premises. Once more the argument establishes the reliability of SP only if SP is in fact reliable. And that still leaves us wondering whether that condition is satisfied.

Another fairly simple argument that quickly falls victim to epistemic circularity is an argument from the (natural world) provenance of our basic belief-forming dispositions, including SP.[5] The most prominent contemporary form of such an argument, perhaps the only currently prominent form, is the idea that our tendencies to form beliefs in accordance with SP, memory, introspection, and inductive and deductive reasoning have been selected by evolution, that this wouldn't have happened unless SP and the other basic doxastic practices confer an adaptive advantage, and that they wouldn't do this unless they were reliable.[6] This is certainly an intuitively plausible line of thought, a plausibility indicated by the following quotations.

---

[5]Note that I am restricting myself here to the factors within the world of nature that influence the acquisition of these dispositions. Later we shall consider the very different argument that God has endowed us with these dispositions, and that this guarantees their reliability.

[6]See Stich 1990, chap. 3, for an assemblage of hints toward such an argument, and for acute criticisms of any such line of reasoning.

Creatures inveterately wrong in their inductions have a pathetic but praiseworthy tendency to die out before reproducing their kind. (Quine 1969, p. 126)

Natural selection guarantees that *most* of an organism's beliefs will be true, *most* of its strategies rational. (Dennett 1981, p. 75)

To spell out the argument a bit more, it presumably runs something like this. Our most fundamental belief-forming dispositions have developed in the course of the evolution of the human species, along with the rest of our innate endowments. But they wouldn't have developed unless they gave us an advantage in the struggle for survival. And belief-forming dispositions couldn't make a positive contribution to our ability to survive unless they produce mostly true beliefs. If they were to yield mostly false beliefs instead, we would frequently be acting on misapprehensions, and if that happened frequently enough, we wouldn't last long. Hence, given that the human species has not only survived but even achieved dominance on the planet, it must be that our basic doxastic practices are reliable.

Despite its obvious plausibility this argument is not immune to criticism. Stephen Stich (1990, chap. 3) subjects it to severe criticism on two points. First, he argues that evolution does not generally, much less always, produce optimally designed systems in surviving species. Second, he argues against the view that natural selection is the only important factor in the development of our basic doxastic tendencies.[7] However, I need not get into the question whether the premises of the argument provide adequate support for the conclusion. The only point I need make here is that the argument is obviously epistemically circular. The evolutionary theory that the argument presupposes and makes use of obviously rests on empirical evidence gathered by reliance on SP. And so even if the argument comes out totally unscathed from attacks like those of Plantinga and Stich, it is still disqualified by virtue of relying for its cogency on the assumption of what it is designed to establish. In the course of our discussion we will find again and again that epistemic circularity rears its unlovely head, sometimes in the most unexpected places.

[7]Stich's target is "inferential strategies" rather than doxastic practices, but his arguments are easily adaptable for my purposes. See also Plantinga 1993, chap. 10, in which there is an all-out assault on the assumption that our doxastic tendencies can be adaptive only if they are reliable.

vi. The Road Ahead

Having warmed up on the simplest arguments for reliability, let's turn to attempts we can take more seriously. These can be divided into empirical and a priori arguments. The latter have often been developed explicitly to avoid epistemic circularity; and if they are wholly a priori they will succeed in that. The empirical arguments are clearly threatened by epistemic circularity. Some of their authors make strenuous efforts to avoid this; with what success will appear in the fullness of time. Though it might seem most natural to continue the above discussion by moving on to the more sophisticated empirical arguments, it will prove most fitting to save them for later.

Chapter 3

A PRIORI ARGUMENTS

i. Theological Arguments

A historically famous attempt to shore up confidence in sense perception is that of Descartes in the *Meditations*. Having produced more than one argument for the existence of God, he goes on in "Meditation VI" to argue that God would be a deceiver if sense perception were not a more or less reliable source of belief. For, he says, we have a very great inclination to accept the beliefs we form perceptually, and we have no resources for correcting this inclination. Since we were put in this situation by God, and since God is no deceiver, we can reasonably conclude that sense perception provides us with a fairly accurate account of the world about us.[1]

This is strictly an a priori argument only if we limit the basis for accepting the existence of God to the ontological argument, for of the three arguments for the existence of God to be found in the *Meditations* only the ontological argument is purely a priori. While the ontological argument, in Descartes' form, hangs solely on an analysis of the idea of God, the two arguments in "Meditation III" have as crucial premises: one, that Descartes has the idea of God; the other, that he exists; and Descartes derives both of these premises from his experience of his own mental states. But, according to Descartes at any rate, that experience is a source for belief formation that is as

[1]The reader will recognize in this a supernaturalistic "provenance argument" analogous to the naturalistic provenance argument from evolutionary theory discussed near the end of the previous chapter. In both cases the claim is that by considering the origin of our belief-forming tendencies we can find ample reason for supposing them to be reliable.

independent of sense perception as rational intuition. Hence these arguments might be said to be "functionally" a priori in a way in which we are interested here. They are acceptable independently of any reliance on sense perception as a source of belief, if Descartes is right about the epistemic relations of introspection and sense perception, and hence their inclusion in a larger argument for the reliability of SP runs no risk of epistemic circularity.

Granting the noncircularity of the argument, does it provide sufficient support for its conclusion? I will take it that the argument from the existence of God on is unexceptionable. If the world depends for its existence and its character on an omnipotent, omniscient, and perfectly good God, and given that we have a strong, indeed irresistible, inclination to form beliefs in accordance with SP, then surely the creator would not have given us such an inclination unless following it were for our good. And since, with only minor exceptions, engaging in a doxastic practice is beneficial to us only if it is reliable, it follows without more ado, that SP is reliable. This means that the cogency of the whole argument depends on whether we are justified in accepting the premise that the world is created and governed by an omnipotent, omniscient, and perfectly good deity. As far as Descartes is concerned, that is a question of whether the arguments he gives suffice to yield this conclusion. More generally, the question of whether we can establish the reliability of SP on the basis of a divine guarantee depends on whether it is possible to justify believing in the existence of such a deity without that justification relying, at some point, on the reliability of SP. I will say a few words on both these matters.

I won't enter here onto a detailed discussion of the Cartesian arguments for the existence of God. Suffice it for present purposes to say that one would be hard pressed to find a defender on the current scene. An exception is Clement Dore (1984), who defends Descartes' version of the ontological argument. Of course, there are notable contemporary defenders of other forms of the ontological argument, for example, Charles Hartshorne, Norman Malcolm, and Alvin Plantinga. But only Dore, to my knowledge, defends Descartes' form. And I can't think of even one philosopher who takes up the gauntlet for the arguments in "Meditation III."

But that leaves us with the more general question whether the "theological argument for the reliability of SP", as we may call it, could be both noncircular and otherwise cogent if it were based on different grounds for the existence of God. If we are to avoid epistemic circularity, those grounds cannot draw on SP at any point. That rules out many familiar arguments for the existence of God, including the teleological argument, which depends on empirical evidence for adaptive features of the world, and any form of the cosmological argument, like Aquinas' argument from motion, that begins from a premise, like 'Something is in motion', that rests on sensory observation. More abstract versions of the cosmological argument that take a starting point as unspecific as 'Something exists' may successfully deflect the charge of relying on SP. Perhaps we can know that something exists by cogito-like maneuvers: even if I am deceived by an all-powerful demon, still I exist or I couldn't be thus deceived. Or perhaps we can know this just by introspection, without depending on any sensory input from the external world. The ontological argument I have already acknowledged to be purely a priori, and some forms of the argument are more impressive than Descartes' version. Moral arguments will fall into the a priori category if they can be based just on my awareness of moral standards plus reflection on what is necessary for that. Thus there are arguments for the existence of God, other than those of Descartes, that do not rely on SP.

But do any of these arguments establish their conclusion? I myself find some of them, in their strongest forms, not wholly lacking in force. I believe that, for example, the cosmological argument as developed by Samuel Clarke (1738) and updated by William Rowe (1975), and the ontological argument as developed by Plantinga (1974), do provide a significant basis for believing in the existence of the kind of deity envisaged in Descartes' argument for the reliability of SP. Moreover, there are other SP-independent grounds for the belief in such a deity that I consider to have some considerable weight, particularly the experience of (what seems to the subject to be) the presence and activity of God.[2] If I were defending the theological argument for the reliability of SP, I would have to defend the claims just made. But that is not in the cards here. Rather, what I want to say is that even though the grounds in question make a significant contribution to the justification of theistic belief, they cannot carry the

[2]See Alston 1991a.

whole load. If theistic belief is to be sufficiently justified for rational acceptance, these SP-innocent grounds will have to be supplemented by others that do rely on SP. They will have to be set in the context of a "cumulative case" for theism that also includes such elements as the following.[3]

(1) A tradition of "encounters" with God, communications from God, and experiences of God, such as we have in, for example, the Christian tradition.

(2) The experience of "payoffs" in leading the life that the religious community represents as enjoined by God, especially where these payoffs can be seen as fulfillments of what the tradition represents as divine promises.

(3) Additional natural theological arguments that depend on SP for some of their premises.

These additional components of the cumulative case do depend on SP. That has already been specified for (3). As for (1), I can't know what the tradition consists of without using my senses to learn this—from reading written records, hearing it proclaimed, discussing it with other people, and so on. I can't get at all that just by rational reflection and introspection. As for (2), I might pile up some experience of leading the life enjoined by a religious community without using my senses (though I doubt this), but I couldn't have any reason to think that this life is what is enjoined by the community except by reading or hearing what representatives of the community have to say. Moreover, evidence of this sort is not very impressive when confined to my own case. It is much stronger if it involves patterns that are exemplified by many members of a group. And, obviously, to find out that this is the case I have to use my senses.

Thus, when all is said and done, a theological argument for the reliability of SP that has the best chance of being cogent will depend on SP for some of its premises and so will, after all, run afoul of epistemic circularity. Even apart from circularity, the cogency of the strongest form of such an argument will be highly controversial. As a theist I believe that the argument, apart from epistemic circularity, is

[3]For the concept of a "cumulative case" for religious belief, see Basil Mitchell 1981 and Alston 1991a, chap. 8.

quite strong. But that one flaw is sufficient to disqualify it from the present competition.

Nontheological a priori arguments on the current scene virtually all stem from Immanuel Kant or Ludwig Wittgenstein or both. The prime Kantian source is the transcendental deduction of the categories in the *Critique of Pure Reason,* more specifically the attempt to show that a necessary condition of the unity of consciousness is that the categories apply to any objects of experience. The Wittgensteinian sources are more diverse; they include the private language argument, the view that concepts have "criteria" for their application, and, more generally, verificationism. I will reverse the historical order and deal first with attempts that are more or less in the spirit of Wittgenstein.

ii. Verificationism

The simplest of the arguments that I am calling Wittgensteinian is a verificationist one that goes as follows.

> If it is a serious question whether SP is reliable, then it must be a meaningful hypothesis that SP is not reliable. It must be conceivable that our sense experience is just as it is even though the things we take it to reveal are not generally the case. But this is not a meaningful supposition, for there is no conceivable way in which it could be empirically confirmed. What empirical evidence could count in favor of it? We can have empirical reasons for correcting sense perception on points of detail, as has happened repeatedly in the history of science from the seventeenth century. But what possible empirical basis could there be for a wholesale rejection of the deliverances of SP? And if there is no possible empirical confirmation for such a view, it is without factual meaning. Hence there is no meaningful alternative to the reliability of SP. We can know, just by considering the conditions of meaningfulness, that SP is reliable.[4]

---

[4]My justification for calling this argument 'Wittgensteinian' is rather indirect. First, there is the influence of Wittgenstein on the Vienna Circle, and this argument is the one that made the Vienna Circle famous. To be sure, it is usually more metaphysical theses like "the existence of the external world" or "realism" that the logical positivists attacked on verificationist grounds. (See, e.g., Moritz Schlick, "The Turning Point in Philosophy" and Rudolf Carnap, "The Elimination of Metaphysics Through the Logical Analysis of Language", both reprinted in Ayer 1959). But the arguments could easily be transferred to the question of the reliability of sense perception. Second, Wittgenstein, despite the disavowals of many of his

I have presented this as an argument for the reliability of SP. Another version, more in keeping with the spirit of verificationism, would present it as an argument for the meaninglessness of both sides of the issue—reliability or unreliability. Since no conceivable empirical evidence could decide the matter, there is no real matter of fact at issue. It is a pseudoquestion. This version doesn't look like a defense of the reliability of SP, but it has the same effect. Since it forbids us to question something we all take for granted in practice, it leaves that conviction in undisturbed possession of the field.

This is designed to be an a priori argument, since it purports to rest only on conceptual considerations as to what it is for a sentence to have "cognitive" or "factual" meaning. The claim that such matters can be settled definitively just by a consideration of concepts is questionable. For that matter, the claim that there is an interesting and important concept expressed by terms like 'cognitive meaning' and 'factual meaning' can be questioned. But I will waive these doubts and allow that the argument counts as a priori.

The most direct way to attack the argument would be to argue against the verifiability criterion of factual meaningfulness, something I am by no means indisposed to do. However, to do that properly is a long and complicated business, and I shall confine myself to the following point, which is sufficient for the purpose at hand. The criterion *presupposes* the by-and-large reliability of sense perception. What would be the point of requiring empirical *verifiability* or *confirmability* of p as a necessary condition of the factual meaningfulness of p, unless it were possible to verify or confirm a hypothesis by relating it properly to the results of observation? And that is possible only if the results of observation are by-and-large correct; otherwise, the fact that a hypothesis leads us to expect results that are observed to obtain is no reason to suppose it to be *true*. Hence the argument turns out to presuppose that which it is invoked to prove.

An antirealist form of verificationism might attempt to disengage the Verifiability Principle from any commitment to what would suffice to *verify*, or *confirm*, a proposition, understanding verification as show-

---

followers, does use verificationist arguments not infrequently. We will see a famous one later in this chapter, the private language argument. Third, this argument is beautifully explicit in Bowsma (1949), a prominent Wittgensteinian. However, nothing hangs on the Wittgensteinian label. You can discard it if you like.

ing the proposition to be true, and confirmation as showing the proposition to be probably true. On this version, empirical testability would just amount to the possibility of comparing the implications of the proposition in question with what we believe on the basis of SP, without any further claim that these latter beliefs are (even generally) true. But in that case, verificationism would not support the reliability of SP as I understand it, since it would not support the thesis that the reliability of SP (in the sense of the tendency of SP to yield truths, in a realist sense of 'truth') cannot be questioned. On the contrary, on this nonrealist interpretation the view would be turning its back on truth, and hence on reliability.

iii. Criteria of Physical Object Concepts

In this and the following two sections, we turn to Wittgensteinian arguments that are based on considerations concerning language. The first of these is based on the idea that the terms of our language, at least terms denoting physical and mental properties and kinds, have *criteria* for their application. A *criterion* for the use of 'P', in this sense of the term, is a basis, ground, or reason for supposing the term to apply that has that status by virtue of the meaning of 'P', or, as Wittgenstein prefers to say, by virtue of the "grammar" of 'P'. If a behavioral criterion for 'is (emotionally) upset' obtains, then we have a good reason for supposing that the person is upset, just by virtue of the concept of being upset, by virtue of what the term 'is (emotionally) upset' means. The satisfaction of the criterion doesn't logically entail that the person in question is upset; it is possible for a person to exhibit the criterial symptoms of being upset and yet not really be in that condition. The person might be shamming, or these manifestations might, atypically, be due to something else in this instance. Nevertheless, if the criterion obtains we are guaranteed a priori, just by the meaning (use) of the term (by the constitution of the concept), to have a good reason for supposing the person to be upset.

Wittgenstein's main application of the notion of criteria was to mental terms and concepts. He was concerned to argue that our most basic third-person indications of what other people are thinking and feeling do not owe their efficacy to empirical correlations between those indications and what they indicate, but that they are guaranteed to be good indications just by virtue of the meanings of our terms for thoughts and feelings. He used this both to combat skepticism about knowledge of other minds, and to oppose the view that each person has a privileged epistemic access to his/her own mental states. He especially strongly opposed the claim that only Jones can really know what Jones is thinking or feeling. But Wittgenstein did not suppose that the notion of criteria applies only to mental terms, and, in any event, his followers have used the notion more widely. All this suggests an a priori way of supporting the epistemic claims of sense perception. If X's looking a certain way is a criterion of its being red, or round, or a tree, then it is logically impossible for its looking that way not to be an adequate ground of X's being red or round or a tree. We can know a priori that X's looking that way is a good reason for supposing it to be red (round, a tree). Remember, too, that we are not taking SP to be confined to basing perceptual beliefs only on sensory appearances but to include input from background beliefs where that is relevant. Hence, more generally, criterion theory suggests the possibility of holding that what, in the use of SP, we employ as the most fundamental grounds of perceptual beliefs—both experiential and doxastic grounds—are criterially related to the beliefs they are grounding. It is by virtue of what is meant by 'That is a Victorian house' that when something looks as that does now (and when other things are as the operative background beliefs, if any, have it), then we have an adequate basis for supposing that to be a Victorian house.

In Wittgenstein's *Philosophical Investigation* (1953), we find a suggestion for an application of criterion theory to the epistemology of perceptual beliefs.

> The fluctuation in grammar between criteria and symptoms makes it looks as if there were nothing at all but symptoms. We say, for example: "Experience teaches that there is rain when the barometer falls, but it also teaches that there is rain when we have certain sensations of wet and cold, or such-and-such visual impressions". In defense of this one says that these sense-impressions can deceive us. But here one fails to reflect that the fact that the false appearance is precisely one of rain is founded on a definition. (#354)

In other words, just by knowing the meaning of 'rain' we can know that such-and-such sensory impressions are a good reason for supposing there to be rain in the vicinity. Here attention is confined to an experiential criterion. But a more inclusive conception of criteria for perceptual beliefs is suggested by John Pollock (1974).[5]

> To learn the meaning of a concept is certainly not to learn its "definition". It is to learn how to use it, which is to learn how to make justifiable assertions involving it. Thus it seems to me inescapable that the meaning of a concept is determined by its justification conditions. (P. 12)

> Just what is necessary before we can truly say of a person that he has learned the concept of a certain kind of thing, such as "red thing" or "bird"? We frequently talk about a person "knowing what a bird is" rather than about his having the concept of a bird. But we must be careful with the locution "S knows what a bird is", because it can be used to mean two quite different things. There is a clear sense in which an ornithologist knows more about what a bird is than does a layman, e.g., he knows that birds are warm blooded, that they have livers, etc. But this sort of knowledge about birds cannot be part of having the concept of a bird. . . . Someone must already have had the concept of a bird before these contingent facts could have been discovered. . . . It is the sense of "S knows what a bird is" in which the naturalist had to already know what a bird was that is equivalent to "S has the concept of a bird". This just means "S knows a bird when he sees one." . . . To say that S knows a bird when he sees one is just to say that he can pick birds out from among other things—he can identify birds. . . . But this is still not as clear as we might desire. What is necessary in order for a person to know how to identify birds? There is perhaps a temptation to say that one must know what would count as making "x is a bird" true—one must know the truth conditions for "x is a bird". But it does not take much reflection to see that this is not the case. Although a philosopher or a lexicographer *might* (although I doubt it) be able to construct a definition of "bird" that would give us such a set of truth conditions, very few ordinary speakers of English would be able to do this. . . . But one might suppose instead that in order for a person to know how to identify birds, although he need not be able to say what makes something a bird, he must nevertheless "do the identifying"—ascribe the concept "bird" to things—just when the truth conditions are satisfied.

[5]Actually, the discussion Pollock gives of perceptual beliefs is almost entirely in terms of experiential bases, with little recognition of the role of background beliefs. However, I am not concerned here to discuss the details of Pollock's epistemology, but only to discuss the bearing on our problem of a general account of criteria that is suggested by passages such as those I quote.

But if this is understood as requiring that the person never make a mistake, then clearly it is too stringent a requirement. . . . If his ascription of the concept were . . . justified, even though false, this would excuse his mistake from showing that he had not learned what a bird is.

But if in fact the child did not *know how* to ascribe the concept and its complement justifiably (i.e., he did not know how to justifiably determine whether something was a bird), this would show that he had not learned how to identify birds and so does not have the concept.

Conversely, when the child has learned to judge justifiably whether a thing is a bird (i.e., he has learned to ascribe the concept and its complement to things justifiably), we are satisfied that he knows how to identify birds and so has got the concept right—he knows what a bird is. (Pp. 13–15)

In other words, to possess the concept of a bird, or of any other kind of thing or property that is perceivable (let's follow Pollock in calling these "perceptual properties") is to have the ability to make justified attributions of it; it is to have a working knowledge of the "justification conditions" of the concept. But if that is what it is to have the concept, then what the concept is can be specified by laying out those justification conditions. Those conditions constitute the content of the concept.

It was assumed that having once spelled out the justification conditions for a statement, we would have to go on to prove that those are the justification conditions by deriving them from the meaning of the statement (which was identified with the truth conditions). To prove that the purported justification conditions are the justification conditions would be to derive them from something deeper. But in fact there is generally nothing deeper. The justification conditions are themselves constitutive of the meaning of the statement. We can no more *prove* that the justification conditions of "That is red" are the justification conditions than we can prove on the basis of something deeper about the meaning of "bachelor" that all bachelors are unmarried. Being unmarried constitutes part of the meaning of "bachelor" and as such cannot be derived from anything deeper about the meaning of "bachelor"; and analogously the justification conditions of "That is red" or "He is in pain" are constitutive of the meanings of those statements and hence cannot be derived from any deeper features of their meanings. There are no deeper features. (P. 21)[6]

---

[6]In fairness to Pollock it should be pointed out that he does not give this account of all concepts. Here we are concerned only with concepts of relatively immediately perceivable things and features of things.

But if concepts of perceptual properties consist of the conditions under which one is justified in perceptually attributing those properties to something, one is thereby in a position to determine just by reflection on the nature of the relevant concepts what it takes to make justified perceptual attribution of those concepts to something. And if we may assume that SP embodies, at least in the main, ways of forming perceptual beliefs that are in accordance with those justification conditions (forming perceptual beliefs, mostly, when and only when the conceptually determined justification conditions are satisfied), we can establish the positive epistemic status of SP just by deriving those justification conditions from the concepts in question. And since this only requires reflecting on the concepts, and does not require any appeal to what we learn from perception, it is a wholly a priori procedure and does not involve us in epistemic circularity.

Before considering what is to be said for and against this position, I will consider what would be its most plausible form. There are two basic questions here. (1) Are the justification conditions for perceptual concepts restricted to sensory appearances? (2) Is it plausible to suppose that all, or any, concepts of perceptual properties consist wholly of justification conditions? I will consider these in turn.

The first question can be discussed at any given length, but I will dispatch it promptly, so as to move on the second, which is more relevant to my present concerns. It seems to me that sensory appearances at least bulk large among the justification conditions for perceptual properties. This is true not only of "sensible quality" concepts, like the concepts of colors, smells, and tastes, but even for concepts of perceivable kinds like dogs, trees, houses, and books. We not only typically, and justifiably, take something to be red, acrid, or sour by the way it looks, smells, or tastes. We also typically, and justifiably, take something to be a dog, a house, or a tree by the way it looks. To be sure, other matters can and do figure in considerations of the justification of perceptual beliefs. Questions can be raised as to whether one's senses are working properly, whether conditions are favorable for accurate observation, whether it's at all likely that a tree would be found here, and so on. But it could still be claimed that the sensory appearance alone is sufficient to render the subject prima facie justified in supposing the object to be red, a book, or whatever. These other matters can then figure as possible overriders of that prima facie justification.[7] At least, I want to grant for the sake of argument that the justification conditions involved in a perceptual property concept could be restricted to sensory appearances, and move on to the second question.

As for that question, it seems to me very dubious whether any concepts of perceptual properties are made up wholly of justification conditions. This is especially obvious if justification conditions are restricted to sensory appearances. I make that judgment even with respect to "sensory qualities" like colors, pitch, loudness, and smells. The simplest version of a concept of the physical property of redness is that it is the property by virtue of having which an object will give rise to visual appearances of redness in a normal observer in normal circumstances.[8] Thus, the concept includes such components as the concept of a property, the concepts of normal observer and normal circumstances, and the concept of a disposition to produce certain results in certain circumstances. Furthermore, there is seepage from scientific results. Even a moderately knowledgeable person might include in her concept of *red* that it has to do with the frequencies of light waves reflected from the surface of the object. Thus it would seem that even here the concept is not made up wholly of sensory justification conditions. I have, of course, been speaking in terms of a realist conception of sensible qualities. But even on a phenomenalist account, in which the concept of redness as a property of physical objects is the concept of conditions under which observers would have certain kinds of visual sensations, the concept includes abstract concepts like "conditions under which".

As for concepts of perceivable kinds (trees, dogs, houses, books), typical sensory appearances are much less prominent. Here we find various technical or scientific concepts where sensory appearance components are not at all central. The biologist's concept of a tree is the concept of something with a certain organic structure that carries out certain vital operations. Anything that satisfies these conditions, however it looks, feels, or smells will be a tree. On the other hand, sensory appearances figure prominently in *common-sense* concepts of perceivable kinds. It is true that not all trees, all dogs, or all houses

---

[7]This issue is discussed in considerable detail in Alston 1991a, chap. 2.
[8]For some powerful reasons for regarding this version to be oversimple, see Hardin 1988.

look just alike. Indeed, the range of looks is quite wide. Nevertheless, there are presumably commonalities in the visual appearances of all (almost all?) trees, dogs, and houses. Moreover it may be that some disjunction of looks finds a place in such concepts. At least we do readily pick up the knack of identifying something as a tree, a dog, or a house by the way it looks. But there is typically much more to such concepts. True, there may be rudimentary concepts of such things that are solely constituted by sensory appearance types. Our earliest concept of a house may be *something that looks sufficiently like our earliest paradigms of houses.* But it is presumably part of even the most modest adult concept of a tree that it is a vegetable organism, of a dog that it is an animal organism, and of a house that it is a shelter for habitation.

In considering whether the concepts in question go beyond justification conditions we have to consider whether some or all of these non-sensory-appearance components figure in as justification conditions. Again without going fully into the matter, it seems clear that not all such components have this status. Do I have to ascertain that the object before me is a vegetable organism, with all that involves, in order to be justified in supposing it to be a tree? Isn't it sufficient that it looks enough like a tree, in the absence of any reason for supposing otherwise? Do I have to determine that the object is disposed to look red to normal observers in normal circumstances in order to be justified in supposing it to be red? Isn't it enough that it looks red and that there be no reason to suppose that there is something in this situation that defeats the normal presumption that what looks red is red? It certainly seems that even if justification conditions are involved in our concepts, they do not exhaust those concepts.

In a later work, Pollock (1986) explicitly recognizes other components of concepts in general, in addition to justification conditions.

What makes a concept the concept that it is is the way it enters into various kinds of reasoning, and that is described by saying how it enters into various kinds of reasons, both conclusive and prima facie. Let us take the *conceptual role* of a concept to consist of (1) the reasons (conclusive or prima facie) for thinking that something exemplifies it or exemplifies its negation, and (2) the conclusions we can justifiably draw (conclusively or prima facie) from the fact that something exemplifies the concept or exemplifies the negation of the concept. My proposal is that concepts are individuated by their conceptual roles. The essence of a concept is to have the conceptual role that it does. (Pp. 147–48)

In the earlier book, (1) was called "justification conditions".

What are we to say of Pollock's position? First, let's ask what reasons there are to accept the claim that concepts of perceptual properties contain justification conditions. Pollock talks as if we can ascertain this by the standard techniques of philosophical analysis— asking ourselves under what conditions one would "have" or "know" the concept, what it would take to show that one lacks the concept, how the concept could be explained, and similar questions about the meanings of words that express the concept. But, like most exciting claims about the analysis of concepts, the deployment of such techniques by acute philosophers fails to achieve anything like consensus. For every eminent thinker who accepts an account like this we can find one who, after careful consideration, rejects it. And so we are motivated to look for some more general reason for favoring the position.

Such reasons are not far to find. First, there is the point that the theory enables us to avoid skepticism about SP and other basic doxastic practices. In the book from which I have been quoting, Pollock introduces this theory of concepts as a way out of skepticism. And so it is. But that is hardly a sufficient recommendation. Pollock presents an eight-step argument for skepticism about "our sources of knowledge"; the conclusion can be avoided by denying, or presenting an alternative to, any of the steps. Why suppose that the premise "The only way to analyze the meaning of a statement is to give its truth conditions" (p. 11)—the one for which this theory of concepts is an alternative—is the one to deny?

At least for certain kinds of concepts, the criterion view has been claimed to be necessary for the very possibility of a public language. Thus, Wittgenstein held that unless concepts of (terms for) mental states had public criteria (in this technical sense of 'criteria'), we could not justifiably make third-person attributions of such terms. And in that case those terms could not acquire an intersubjectively shared meaning in a public language. For in first-language learning, in order for me to learn for what kind of mental states speakers of the lan-

guage use the term 'feel relieved', I have to be able to correlate
something with uses of this term by other people, when they are using
the term of still other people as well as when they are using it of me.
But I can't correlate it with others' feelings; they are not available to
me for that purpose. I can correlate it only with what is publicly
observable; and unless that is tied by the meaning of the word to the
kind of feeling in question, I will not thereby acquire a meaning of the
word by virtue of which it denotes cases of feeling relieved. An
analogous argument can be given for terms for perceptual properties. Unless I can correlate fluent speakers' use of such a term with
something that is both available to me when in the first-language
learning situation (my sensory experiences) and also semantically tied
to the denotation of the term, I will never be able to acquire the
standard meaning of the term in the language.

Concentrating on the latter argument concerning perceptual
properties, it rests, inter alia, on a crucial epistemological assumption
about the conditions under which I can know what is the case in the
external environment. It assumes that unless something directly accessible to me is conceptually tied to such matters, I will be forever
enclosed within the bounds of my own experience. But this can be,
and has been, questioned. Any direct realist will suppose that we can
get perceptual knowledge of various facts concerning the physical
environment even if our thought and language concerning such facts
is not conceptually tied to features of our experience. One way of
spelling out such a position would be in terms of an innate reliable
tendency to form rudimentary beliefs about the external world upon
experiencing certain sensory appearances. More complicated tendencies could then be built up on the basis of experience.[9] There are
also various alternatives to the epistemology underlying the first argument (concerning mental terms), including not only the familiar
"argument from analogy" for conclusions about the mental states of
others, but also the postulation of innate tendencies to form beliefs
about the mental states of others on perceiving certain public states of
affairs. An "inference to the best explanation" is still another possibility in both cases. Thus I would say that before we are in a position
to accept the claim that the truth of the criterion view is a necessary

[9]See, e.g., Reid 1970, chap. 4, secs. 21–24.

condition for the possibility of a public language, we would have to eliminate all the alternatives to the epistemology presupposed by the argument for that claim. So far as I know, this has not been done, and the prospects do not seem rosy.

Now for the opposition to the criterion view. Here is a reason for denying that our most primitive physical object concepts embody justification conditions. If I am to have a concept that features such conditions, I must have the concept (a concept) of epistemic justification. Otherwise it could be no part of my concept of a tree that I am *justified* in supposing a tree to be before me when such and such conditions are satisfied. Perhaps this requirement of a concept of epistemic justification could be minimally satisfied without my being able to explicitly entertain propositions to the effect that a given belief is or is not justified, and without my having any ability to verbally express such a concept. But whatever is required for the minimal possession and use of a concept, I must satisfy that with respect to the concept of epistemic justification if justification conditions are to be a part of any concept of mine. And it seems clear that the least sophisticated humans, very small children for example, lack any concept of justification even in its most rudimentary form. More specifically, it seems clear that very small children acquire concepts of kinds of perceivable items in their physical and social environments, and of the perceivable properties of such items, well before they can be credited with even the most rudimentary grasp of a concept of epistemic justification. Hence we can hardly suppose that justification conditions form part of *any* physical object concepts possessed by such subjects. A two-year-old knows what doors and windows, birds and dogs, adults and children look like, and can recognize them perceptually; but it would be rash to suppose that the two-year-old can wield a concept of being epistemically justified in a belief.

Once we recognize that one can have a usable concept of a dog or a window or a person without any justification conditions figuring in that concept, one is led to wonder whether the presence of such conditions in more sophisticated concepts, if that is the case, has the epistemological consequences Pollock supposes. Remember that Pollock is arguing that it is conceptually impossible for us to fail to be justified in believing that there is a tree in front of us when, as we might put it, the standard justification conditions are satisfied, since the sufficiency of those conditions for justification is carried by the very concepts we use in forming the belief in question. But even if that is true with respect to typical adult concepts of trees and the like, it would appear to be always open to us to retreat to a more primitive concept of a tree that embodies no such justification conditions.[10] Once we see that possibility, the question arises as to why we should use the more sophisticated rather than the less sophisticated concept. And that, at bottom, is just the question of whether the justification conditions embodied, by hypothesis, in the typical adult concept of a tree, are in fact sufficient for justification of perceptual beliefs that a tree is present. Hence, we have in no way evaded or resolved that question by pointing out that our typical adult concepts embody an answer to it. We can still ask whether we are well advised to use those richer concepts rather than their more primitive analogues. And that is equivalent to the question of whether the epistemological commitments involved in those richer concepts are correct, valid, acceptable, or justified. This argument, I take it, shows that even if our mature concepts of perceivable properties do embody justification conditions, pointing this out does not suffice to show that these are the correct justification conditions. Thus, even if "criteria" or "justification conditions" are built into typical adult concepts of perceptual objects and properties, there are strong reasons for doubting that this has the epistemological consequences the likes of Pollock take it to have.

Moreover, one of the points we made earlier has a similar bearing on the case. Remember the point that we cannot plausibly suppose that concepts of perceptual properties are, in general, made up solely of justification conditions. This casts doubt on the possibility of a conceptual support for the justificatory efficacy of SP, even if concepts of perceptual properties do contain justification conditions. Once we recognize that there is more to the concept than those

---

[10]Pollock (1974) argues that since concepts do not, in general, consist of truth conditions, the only alternative is that they consist of justification conditions. But he works with an excessively narrow conception of truth conditions, according to which a concept is given by truth conditions only when we can spell out a noncircular verbal definition that specifies necessary and sufficient conditions for truth. But there is really a wider range of possibilities before us, including concepts the truth conditions of which can be verbally indicated only by something as loose as 'whatever is sufficiently like such and such paradigms'. As this example indicates, I also recognize concepts that exhibit vagueness, open texture, and other kinds of indeterminacy.

justification conditions, and hence that there is extra content to the belief that those conditions are said to justify, we are ineluctably faced with the question as to whether the concept is "well advised" to conjoin that extra content with those conditions of justification. For the sake of a simple example, let's suppose that the justification conditions contained in the concept of a tree simply concerns typical tree-looks, but that the concept also contains such components as *vegetable organism* and *circulation of sap*. Then we are faced with the question, "Why should we suppose that one of these tree-looks is an adequate basis for supposing that what looks that way is a vegetable organism and involves circulation of sap?" It won't do to reply, "It is conceptually necessary that the application of a concept with that content is justified by that kind of look", for the question is as to the validity, justifiability, or rationality of *using* a concept that is structured in this way. If the mere fact that a concept includes certain justification conditions renders applying that concept rationally acceptable on the basis of the satisfaction of those conditions, then we could justifiably apply any conceptual content on any basis whatever. Simply adjoin to the actual content of the concept of 'electromagnetic field', 'capitalist class structure', or 'divine creation' the justification condition *looks like a crow*, and we have a conceptually guaranteed way of justifiably saying of something that it is an electromagnetic field, etc.[11] It can't be (shouldn't be) that easy. No doubt, someone will point out that our familiar concept of a tree grew up naturally; it wasn't artificially constructed like the examples I just gave. And that may well be a crucial difference. But then the difference is not that the status of the justification conditions is guaranteed by the concept in the one case and not in the other. The difference is rather in how the concept originated and what place it has in our customary practices. That is another matter altogether, a matter to which I will return in the last chapter, in which I offer my reaction to the impossibility of a cogent noncircular demonstration of the reliability of our familiar doxastic practices.

Thus, even if it is a fact that a given concept embodies "justifica-

---

[11]The reader will note the similarity of this point to the previous point that if there are rudimentary concepts of perceptual properties and kinds that do not embody justification conditions, we are faced with the question as to why we should use the more developed concepts that feature justification conditions rather than the embryonic concepts without them.

tion conditions", that does not suffice to give a conceptual, a priori guarantee of the suitability of those conditions, their adequacy to the justification of the application of the concept in question. And so, even if concepts of perceptual properties do consist, in part, of justification conditions, that doesn't really enable us to show, by conceptual analysis, that the satisfaction of those conditions justifies us in taking the whole concept to apply.[12]

Let's take stock. We have seen that it is dubious that perceptual concepts do, generally, contain justification conditions. And we have seen that even if they do, that fails to provide a conceptual guarantee for the justificatory efficacy of our standard ways of forming perceptual beliefs. But now I want to assume, for the sake of argument, that there are strong reasons for adopting some kind of "justification conditions" account of perceptual concepts of perceptual properties, *and* that this provides an a priori justification for claims as to the conditions under which perceptual beliefs are justified. And I will then argue that even so, we lack a cogent noncircular argument for the *reliability* of SP.

In following the development of a Wittgensteinian theory of criteria, we have lost sight of our central problem concerning reliability. Wittgenstein and his followers on this point, including Pollock, aspire only to show that the constitution of the relevant concepts guarantees that our ordinary perceptual beliefs are *justified*, or that their perceptual grounds constitute *good reasons* or *adequate bases or grounds* for

---

[12]A desperate response to this criticism would be to grasp the nettle and say: "Well, that's true of our ordinary concepts of trees and books, of ships and sealing wax. We are hopelessly enmeshed in skepticism so far as they are concerned. But that just shows that our aspiration level for perceptual justification was too high. We can't show that we are perceptually justified in applying those ordinary concepts. So let's perform radical surgery. Let's cut these concepts down to a working size. Let's remove everything other than their ordinary justification conditions, the ones that we standardly take as sufficient grounds in SP. Then we can show just by conceptual analysis that SP yields justified beliefs." But this is to gain a Pyrrhic victory. The concept of a tree now becomes: *whatever can be justifiably asserted to be exemplified on the basis of A, B, C . . .* , where A, B, C . . . are the justification conditions in question. This is just a special kind of reductive approach to answering skepticism, where the skeptical allegation of a gap between our grounds for a belief and the content of the belief is met by cutting the latter down to the measure of the former. That can always be done. At least it can always be done if the surgery leaves us with a coherent concept. But it "solves" the problem only by changing the subject. It is now conceptually guaranteed that SP provides us justified beliefs of the form, 'That's a tree'; but that is only because 'That's a tree' now means something like 'That's a thing of the sort for the identification of which certain perceptual grounds provides sufficient grounds'. And that just isn't what we are interested in when we wonder about the epistemic status of SP. We are still left with the question: Does SP provide adequate grounds for perceptual beliefs as they are ordinarily understood?

them. And it would seem that they do not understand those positive epistemic statuses in such a way that they entail reliability. Wittgenstein is generally too slippery to pin down on a point like this, but Pollock is quite explicit about the matter.[13] And he is certainly well advised in taking this position. How could the fact that a belief that a tree is before me satisfies requirements for justification that are built into the concept of a tree guarantee the *truth* of the belief? Any sense of 'truth' in which I could guarantee truth just by building justification conditions into my concepts would be a wildly nonrealistic conception that I would not dream of taking seriously, nor does Pollock. It can't be that easy to bring it about that our thoughts conform to the way things are. Hence no concept of justification that is such that what it takes for justification depends on the constitution of our concepts can carry any implications for truth or reliability. And so this whole discussion of the Pollockian position and, more generally, of Wittgensteinian criteria has been beside the point, so far as our central concern with reliability is concerned. I could have pointed out at the beginning of this section that the Wittgensteinian position, including Pollock's version, is intended to provide a conceptual, a priori validation of the *justification* or *rationality* of our ordinary beliefs, including perceptual beliefs, rather than their by-and-large truth. I have taken you, dear reader, on this trip, because of the prominence of the Wittgensteinian notion of *criterion* in discussions of the epistemic status of perceptual and other beliefs. Though in the end our judgment is that the whole program is irrelevant, it was still worth our while to see the many complex issues that are involved in assessing this point of view in its own terms.

iv. Paradigm Case Arguments

We have been exploring the view that terms for (concepts of) perceivable features of the environment carry perceptual justification conditions as part of their meaning. We saw that forbidding obstacles loom in the path of one who would exploit this alleged fact to provide an a priori demonstration of the suitability of what are commonly taken to be sufficient conditions for the *justification* of perceptual

---

[13]See Pollock's rejection of externalist epistemological theories (1986, chap. 5, sec. 4).

beliefs. We further saw that even if that project should succeed, it would still not show that SP is *reliable;* and it is the latter issue with which we are concerned here. Now we shall look at the even more deeply Witgensteinian idea that, by the very nature of language, it is necessary that our ordinary ways of forming and validating perceptual beliefs yield mostly true beliefs.[14] The Wittgensteinian lines of argument to be considered now differ from the criterion argument in more than one way. First, they explicitly start from very general considerations of what is necessary for language, or for having, knowing or using a language, rather than from claims about the meanings of certain kinds of terms or the constitution of certain kinds of concepts. Second, they seek to establish conclusions concerning truth and/or reliability, rather than justification or rationality. Hence they promise to have a more direct bearing on our problem.

First I will consider a form of what has come to be called the 'paradigm case argument'. The general idea of such arguments is that since we learn the meanings of terms by reference to paradigm cases of what they denote, they couldn't have the meaning they do unless those putative cases were genuine. Our grasp of the meaning of 'dog' presupposes that the cases of dogs used to teach us what 'dog' means really are dogs. What we mean by dog is roughly 'something similar enough to x, y, z . . .', where x, y, and z are socially accepted paradigm cases of dogs, and where the 'similar enough' part can be spelled out in various ways. Since our access to the paradigms for such terms is perceptual, the argument, if successful, would show that perception does reliably present us with instances of what we believe ourselves to be perceiving, at least in these cases.

Since I do not wish to enter onto the laborious project of extracting something like this argument from the Wittgensteinian corpus, I will work with a version, put forward as distinctively Wittgensteinian, by Oldenquist (1971).

> Could everybody be fooled all of the time in their judgments of what is red? . . . I think that this is a conceptual impossibility. If *all* English users were taught 'red' . . . by examples all of which were white things illuminated in red light, and no one ever saw what we now call

[14]In our discussion of the criterion view, we saw that one way to defend it was to claim that it is required by the very nature of language. But the view itself does not embody any claims about the essence of language or about what is necessary for any language.

red things, then it would be a contingent truth that red things were white things illuminated in red light.

The defense of this anti-skeptical claim lies in showing that it is conceptually impossible that all of the examples by means of which we all learn the meaning of 'red' should be false examples, and that there-fore it is a conceptual truth that many of our color judgments are actually true. In learning the word 'red' we are taught that *this* and *that* are red. It makes sense for me to say that a learning example is a false example only if I believe it deviates from the general run of such examples. It is the language game played with the teaching examples that defines . . . what it is to be red. To say something is red is to say it is like the general run of teaching examples, regardless of what features we might later discover or hypothesize the teaching examples to pos-sess. (Pp. 411–12)

Note that Oldenquist cannot suppose that his argument just con-cerns how we human beings in fact come to learn the meaning of observation terms and that it has no implications for *what* we thereby come to understand, for what the meaning *is* . If the argument had no such implications, it would always be logically and conceptually possi-ble that the meanings of the terms in question could be learned in some other way that did not involve the presentation of samples, or that our knowledge of the meaning is innate. Oldenquist makes it explicit that he supposes there to be a (somewhat indefinite) refer-ence to the teaching samples in the meaning that is acquired and that is subsequently used. "To say something is red is to say it is like the general run of teaching examples."

In response to this argument I could engage in various quibbles about the ostensive teaching of terms, but I don't feel that to be the heart of the matter. I have no doubt but that we do learn at least the simpler and more basic observation terms in this way, and that a reference to paradigms does play a role in their meaning. Even so, this argument does not show that SP is reliable.

First, let's ask just what range of perceptual judgments its conclu-sion claims to be (mostly) true. Judgments about the teaching sample, obviously. The argument is that since terms like 'red' are taught and learned ostensively, and hence since 'red' means 'sufficiently like m, n, o' (the ostended cases), we can't be mistaken in supposing m, n, and o to be red. For to say that they are red is just to say that they are sufficiently like themselves. But that falls far short of showing that perceptual judgments of redness are generally true, and that is what we would need to establish *reliability*. Reliability requires more than that there be some true outputs; it requires that, in a suitable sample, there be a preponderance of true outputs. And this argument comes nowhere near to showing that. After all, occasions of teaching or learning the meaning of an observation term 'P' will typically constitute a tiny minority of the occasions on which one forms a perceptual belief of some x that it is P. Hence, even if all or most of the attributions involved in learning 'P' are correct, it doesn't follow that most attributions of 'P' are. Why isn't it possible for a person who has acquired the meaning of the term in the way specified to do a poor job of utilizing that semantic knowledge in identifying subsequent examples? This may be unlikely, especially with terms for sensory qualities, but why suppose that it is conceptually impossible? And with respect to kind terms and more complex observable properties, such a pattern may not be terribly uncommon. Hilary Putnam claims not to be able to distinguish elms from beeches. Even if one has undergone a standard course of ostensive teaching one could well wind up in this position. My opponent might claim that this just shows that he has forgotten the meaning of the term, even if he once knew it. But there are a number of other possibilities. One's sensory powers may not be sufficient for the required discriminations in many instances. One may be unattentive and uninterested. One may be mostly confronted with situations in which identification is specially difficult. So why suppose that any argument from teaching samples could establish general reliability?

Oldenquist is aware of this objection. What he says in favor of extending his conclusion beyond the teaching samples is the following. "Wittgenstein's view is, I believe, that something like the learning process, or reinforcement of it, continues throughout one's life" (p. 414). But even if something like this is true, it still seems clear that initial teaching plus subsequent reinforcement occasions constitute only a small percentage of the total spread of attributions. Moreover, Oldenquist's supposition that the sample included in the meaning I attach to 'red' includes all those involved in subsequent reinforcements seems highly implausible.

Once we see that the paradigm case argument shows only that the perceptual beliefs involved in ostension itself are correct, we can see that the argument falls into something like epistemic circularity. For it presupposes something very close to that conclusion. Remember that the argument essentially depends on the claim that the meaning of a perceptual property term can be spelled out somewhat as follows: 'sufficiently similar to m, n, o . . . ' (the teaching examples). But something can be importantly similar to all the teaching examples only if they are importantly similar to each other, only if they exhibit some marked commonality. The supposition that we can successfully impart usable meanings for perceptual property terms by exhibiting a large number of perceptible cases *presupposes* that SP is accurate enough to group together for this purpose (under a common term) only items that are sufficiently similar to each other to yield a usable single meaning of the above form. That can be seen to be very close to the conclusion of the argument, once we see that the conclusion (that all or most of the teaching sample are red things) amounts to the view that all or most of the teaching sample is sufficiently similar to m, n, o . . . For since the teaching sample just *is* m, n, o . . . , any item in that sample will obviously be similar to the group, provided the group as a whole presents a genuine target for the similarity relation. Hence what is presupposed by the argument amounts to the kind of reliability of SP that the argument aspires to establish. In presupposing that our perceptual apparatus can be relied on to group together sufficiently similar items in a teaching sample, we are, as near as makes no difference, presupposing that each item in the teaching sample is sufficiently similar to the members of the group generally. Epistemic circularity has once again reared its ugly head.

Putnam (1981, chap. 1) offers a somewhat similar defense of the reliability of SP, against a skeptical claim that my sensory experience is compatible with my being a brain in a vat and failing to perceive the things I take myself to perceive. The argument appeals to the principle that a term can refer to Ps only if my use of it stands in some real causal connection with Ps. Assuming that the brain in the vat has always been such, its use of 'desk' has never stood in any real causal connection with desks and hence does not refer to desks. The same, according to Putnam, holds of all other terms used in the expression of perceptual beliefs. Hence, the brain in the vat does not have false perceptual beliefs of the same sort as we; rather, it has quite different perceptual beliefs, all or most of which might be true. (The terms in these beliefs might refer to brain states or mental images or what-ever.) And hence there is no possibility that we, meaning what we do by the terms in our perceptual judgments, might be brains in vats, out of any real contact with the external environment. But this argument obviously presupposes that we do mean by our perceptual terms what we would mean if we were in genuine perceptual cognitive contact with the sorts of things we think ourselves to be. And if sense percep-tion is not reliable, we are mistaken about that and we are, seman-tically, in a position more like that of the brain in the vat. Hence, the argument shows that *if* perception is as we ordinarily suppose it to be and is therefore reasonably reliable, we couldn't mean what we do without sense perception's being that way. And that, of course, falls far short of showing that sense perception *is* that way, or that it *is* reliable. Once again the argument, as an argument for the reliability of SP, turns out to be circular.

Sydney Shoemaker (1963) presents something like a paradigm case argument for the thesis that it is necessary that sincere percep-tual statements are generally true, but he focuses on how someone else could tell what I mean by a given term rather than on what it would take for me to learn the meaning of the term. (His term 'perceptual statement' applies to statements to the effect that some-one perceives something in some way, e.g., sees it; not, as is my practice, to statements about the environment that are made on the basis of perception.) The argument runs as follows.

> A primary criterion for determining whether a person understands the meanings of such terms as "see" and "remember" is whether under optimum conditions the confident claims that he makes by the use of these words are generally true. If most of a person's apparent percep-tual and memory claims turned out to be false, this would show, not that the person had exceptionally poor eyesight or an exceptionally bad memory, but that he did not understand, had not correctly grasped, the meanings of the words he was uttering, or was not using them with their established meanings, i.e., was not using them to express the perceptual and memory claims they appear to express. So to suppose that it is only a contingent fact, which could be otherwise, that confident perceptual and memory statements are generally true is to suppose that we have no way of telling whether a person understands the use of words like 'see' and 'remember', or means by them what others mean by them, that we can never have any good reason for regarding any

utterance made by another person as a perceptual or memory statement, and that we could therefore never discover the supposedly contingent fact that perceptual and memory statements are generally true.
And this is a logically absurd supposition. (Pp. 231–32)

But why should we suppose it to be a "logically absurd supposition"? It seems clear that if sense perception is not reliable, then we do
have no way of telling whether another person "understands the use
of words like 'see' and 'remember' ", and "that we can never have any
good reason for regarding any utterance made by another person as a
perceptual or memory statement". How could we determine these
things except by reliance on what we hear them saying and see them
doing, against the background of the rest of what we learn about the
world through perception? But then the denial that we can't determine these things is tantamount to the assertion that sense perception
is reliable. And the claim that it is "logically absurd" that we are in
such a state of deprivation is tantamount to the claim that it is necessary that sense perception is reliable (that sincere perceptual judgments are generally true). But that is just what was to be proved.
Hence the argument is again infected with at least epistemic circularity.

Nor is this an end to the troubles of the argument. Let's ask
Shoemaker why he supposes that "a primary criterion for determining whether a person understands the meanings of such terms as 'see'
and 'remember' is whether under optimum conditions the confident
claims that he makes by the use of these words are generally true". He
may reply, along Davidsonian lines, that the only way we have to
interpret the language of another is by assuming that his statements
are mostly correct. Without making that assumption we could get no
purchase on what he means by any particular linguistic unit. But that
general principle will not yield the conclusion that statements as to
what a person sees, hears, etc., are mostly correct. The assumption
that most of his statements are correct is compatible with the assumption that most claims to be perceiving something are false. For example, it is conceivable (though hardly likely!) that most statements he
makes about the environment on the basis of perception are correct,
while statements to the effect that he is perceiving something are
mostly false. In that case the Davidsonian principle for interpreting

the language is satisfied, while it is not the case that what Shoemaker calls "perceptual statements" are mostly true. We need a reason for holding that this particular minute part of the lexicon—perceptual verbs—could not be interpretable unless statements involving them are mostly correct. And what could that reason be? It is, in general, possible to interpret certain expressions even if statements involving them are not mostly true, provided we have sufficient clues from our understanding of the rest of the language and don't have to rely on the correlation of *these* statements with the facts they are used to assert. Are there such connections of perceptual statements with the rest of the language? Surely there are. We could, for example, by suitable questioning, elicit from the subject statements like "When I see that tree it looks tall and green to me" and "When I hear you talking your voice sounds deep". With enough data of this and other sorts, it would be very plausible to give the normal interpretation to the perceptual terms 'see', 'looks', 'hear', 'sounds', etc., without any assumption of the general correctness of reports of having seen or heard so-and-so.

Suppose we give Shoemaker a break and allow him to broader his category of "perceptual statements" to include statements about the environment based on sense perception as well as explicit claims to perception. In that case, he might be able to argue that perceptual statements are so fundamental to our system of empirical belief that unless they are mostly true our empirical beliefs as a whole are not mostly true. That is surely plausible. And it could be just as plausibly contended that I can ascertain this just by reflecting on my system of belief; I don't need any reliance on the outputs of sense perception, any assumption that those outputs are generally correct, in order to carry this out. Nevertheless, this reflection will tell me only about my own system of belief. What basis do I have for supposing that other people are in a like condition? More specifically, what basis do I have for supposing that a person whose language I am trying to understand is in a like condition? It seems clear that whatever basis I have for supposing that it is true of people generally that their statements could not be mostly true without their perceptual statements being mostly true is gained by my experience of people in their interactions with their world; and this involves reliance on sense perception for my information. Unless I could depend on the reliability of sense perception, I would be in no position to assert the fundamental position of sense perception for people generally. Hence, epistemic circularity has once more made an appearance, at the core of an argument that appears for all the world to be purely a priori in character.

v. The Private Language Argument

Now to get at the very heart of the later Wittgenstein, let's consider the bearing on our problem of the famous "private language argument". This label is misleading in that it suggests that there is a unique argument properly so called. In fact, one can find in the *Philosophical Investigations* a number of lines of argument to which this label can be applied. To add to the confusion, some true believers stoutly deny that there is any such *argument* at all, since Wittgenstein was opposed on principle to propounding philosophical arguments. However that may be, I shall simplify the present discussion by leaving aside any arguments there may be and simply consider what bearing the Wittgensteinian denial of private languages has on our problem. But it will be useful to make explicit what I take to be Wittgenstein's central reason for his denial. It is that language essentially depends on public rules for the use of expressions, rules that are public in the sense that in principle there are ways available to all the members of a community for determining when the rules are being followed or violated. Without such rules no expression has the kind of *use* that makes it possible to employ it to say something meaningful.[15]

First, we must get straight as to just what the anti-private language thesis amounts to. What Wittgenstein denies is not the possibility of a de facto private language, one that in fact is understood and used by only one person, but rather the possibility of a language that is necessarily private, one that only one person *could* understand.

> But could we also imagine a language in which a person could write down or give vocal expression to his inner experiences—his feelings, moods, and the rest—for his private use?—Well, can't we do so in our ordinary language?—But that is not what I mean. The individual

[15]See, e.g., Wittgenstein 1953, #258–70.

words of this language are to refer to what can only be known to the
person speaking; to his immediate private sensations. So another person cannot understand the language. (1953; #243)

If it is impossible for another person to understand the alleged language, then there can be no publicly available system of rules for the
use of the expressions of the language; for if there were such a
system, it would be possible for other people to be acquainted with it,
in which case they would understand the language.

Suppose there can be no private language in this sense. What
bearing does that have on our problem? Let's begin by setting aside
some false scents. It is often said that the notion of a (necessarily)
private language is presupposed by solipsism, by phenomenalism,
and by any attempt to infer the existence of an external world from
our private sensations.[16] For the solipsist supposes herself able to
speak meaningfully even if there is no language community by relation to which her utterances have meaning. And the phenomenalist,
as well as the would-be inferrer of the external world from private
experience, supposes himself to be able to talk about his experiences
prior to carrying out the construction of the public physical world or
the inference thereto. I myself do not agree that the possibility of a
necessarily private language is presupposed by these programs; the
presupposition of a language that it is *possible* for only one person to
understand would seem to be quite sufficient. However, I do not want
to pursue that here. My point is that even if we agree with Wittgenstein on the unacceptability of solipsism, phenomenalism, and the
inference of the external world from private experience, that fails to
show that sense perception is reliable or that the hypothesis of its
unreliability can be dismissed. I take this to be clear on the face of it
with respect to phenomenalism and the inferential program. If those
are dismissed, we are still left with the question of the reliability of
sense perception, a question their dismissal quite fails to answer. The
dismissal of solipsism might seem to have more bearing, since it is
often supposed that solipsism is true if and only if sense perception is
unreliable. But that is not so; the 'only if' is quite justified but not the
'if'. Sense perception could be an unreliable source of belief even if I
am not alone in the universe. In fact, there could be a complex
physical world, even if sense perception fails to provide us with a

[16]See, e.g., Malcolm 1963, "Wittgenstein's *Philosophical Investigations*".

reliable cognitive access thereto. In that case, either we know nothing about it or we have some other basic source of information instead. But the mere rejection of solipsism doesn't enable us to choose between these possibilities.

There is a line of argument, however, that is more relevant to our problem. In setting out my own version of this I will be substantially following Peter van Inwagen in some unpublished comments on a presentation of my paper, "Religious Experience and Religious Belief" (1982), at a symposium of the Central Division of the American Philosophical Association, though the exact shape of the presentation is mine, and van Inwagen should not be held responsible.

Let's use the term 'public language' to cover a language that is used in common by members of a social group (larger than one), a language the terms of which mean what they do by virtue of public rules for their use. Now the Wittgensteinian position on private language could be put as follows:

(1) If (alleged) term 'P' cannot figure in a public language it has no meaning.

But:

(2) If sense perception is not reliable there can be no public language.

The reason for this is that a public language gets established by way of social interactions in which the participants find out by perception what other participants are saying and doing. Think of this in terms of a neophyte. If this person is to become a functioning user of the language, she must be able to get reliable perceptual information about the linguistic and other behavior of her fellow group members. Otherwise she would be able neither to learn the language (how could she?) nor to use it in communication.

But then, by hypothetical syllogism from (1) and (2):

(3) If sense perception is not reliable, no term can have a meaning.

But in raising the issue of the reliability of sense perception, we suppose ourselves to be using language meaningfully. And if we are not using language meaningfully we have failed to raise that issue, whatever we may suppose.

Hence:

(4) If no term can have a meaning, we cannot raise the issue as to the reliability of sense perception.

Therefore, by transposition from (4):

(5) If it is possible to raise the issue of the reliability of sense perception, then terms can have meaning.

By transposition from (3):

(6) If terms can have meaning, then sense perception is reliable.

And from (5) and (6), by hypothetical syllogism:

(7) If it is possible to raise the issue of the reliability of sense perception, then sense perception is reliable.

Thus, there is no real possibility that sense perception is not reliable. If it were not, then there could not so much as be a question of its reliability. If there is such a question, it can have only a favorable answer.

Note that this argument exhibits more clearly a feature of all the arguments we have called 'Wittgensteinian'. They all contend that the hypothesis that sense perception is not reliable is self-defeating. If we try to maintain it, it crumbles beneath our fingers. Either it is meaningless because not empirically confirmable, or it takes away criterial features of the terms we use to formulate particular perceptual judgments, or it would render impossible our learning the meaning of such terms or for such terms to mean what they do, or, as in the present argument, it denies a necessary condition of any term's having meaning.

Though this is, so far as I can see, the strongest argument for the reliability of sense perception (or for the impossibility of supposing sense perception to be unreliable) from an anti-private language perspective, it, too, founders on epistemic circularity. This pops up in the support for premise (2). That support consists in pointing out ways in which we have to rely on sense perception in learning a public language. How else can we learn a public language except by getting reliable perceptual information about the linguistic behavior of speakers of the language as that fits into the physical and social environment? But how do we know that this is the only way? How do we know that we do not have some other cognitive access to the linguistic behavior of others and to the setting in which that is carried on? For that matter, how do we know that our mastery of a public language is not innate, in which case it would not have to be learned at all? Obviously, we know all this because of what we have learned about the world, more specifically because of what we have learned about human beings and their resources for learning, for knowledge acquisition, and for linguistic communication. That gives us good reason for denying that human beings can acquire and use a public language without heavy reliance on sense perception. But we have learned this *by relying on our perception of each other in our physical and social environment and by reasoning from that perceptually generated information.* It is not as if we can know a priori that we have no other cognitive access to these matters. Quite the contrary. A priori it seems quite possible that the requisite knowledge should be innate, or that we should have other, non-sensory modes of access to the linguistic behavior of others and to the stage setting of that behavior. Hence we are relying on sense perception in arguing that a public language presupposes the reliability of sense perception. Epistemic circularity has once more vitiated what looked like a purely a priori argument.

vi. Transcendental Arguments

The next item on the agenda is "transcendental arguments". This label has been much in vogue of late, though as pointed out by Jonathan Bennett (1979, pp. 50–51), it has been variously employed. The contemporary use of the term stems from Kant. In *The Critique of Pure Reason* (1929) he wrote: "The term 'transcendental' . . . signifies such knowledge as concerns the a priori possibility of knowledge, or its a priori employment" (A56; B80). And again: "The explanation of the manner in which concepts can thus relate a priori to objects I entitle their transcendental deduction" (A 85; B117). More specifically, Kant's famous transcendental deduction of the categories sought to show that their applicability to objects of experience is necessary for any experience of objects, since it is necessary for that experience to be attributed to a subject, which in turn is necessary for an experience of objects. We might think of the contemporary use of 'transcendental argument' as broadening out from that Kantian argument in various ways: extending to any argument designed to show that the applicability of certain concepts is required for experience of an objective world, or for self-consciousness, or for experience of the past, or for having any experience whatsoever. The term has also been stretched to cover arguments to the effect that something or other is necessary for the possibility of language or judgment or the possession of any concepts. Thus the label would be applied by some to all or most of the Wittgensteinian arguments we have been considering. Again, since Kant's transcendental deduction was directed at Humean skepticism, the label has been applied to any argument that seeks to show that something the skeptic acknowledges suffices to undermine his skepticism. For present purposes, I shall stick fairly closely to Kant and restrict myself to arguments designed to show that the applicability of certain basic, physical object concepts is a necessary presupposition of any possession of experience by a subject.

Fortunately, the consideration can be brief, since even if the Kantian argument, or any like it in the specified respect, is wholly successful in its announced aim, it will fail to establish the reliability of sense perception. What such arguments are designed to show is that certain fundamental concepts, such as *causality* and *substance* , are applicable to the objects of sense experience. But even if that is so, the question still remains as to whether perceptual beliefs are mostly true. After all, what we come to believe on the basis of perception is almost always much more specific than that we are confronted with substances that are interrelated causally. Hence, even if perception is right in representing its objects as substantial and causally connected (and subject to various other categories), it could be wrong in most of the (alleged) information it provides.

This judgment of irrelevance to our problem applies not only to Kant's transcendental deduction of the categories, but also to the (more or less) Kant-inspired arguments for the applicability of concepts that we find in works by P. F. Strawson (1966, chap. 2, pt. 2), by Bennett (1979), and by A. C. Grayling (1985, chap. 2). The same is to be said of other well known arguments of Strawson's that have been called 'transcendental', such as the argument that one can attribute conscious states to oneself only if one can attribute them to others (1959, chap. 3). Here too we can derive from the conclusion no positive epistemic evaluation of any particular way of forming beliefs about the conscious states of others. Strawson does indeed argue that we have the other-attribution capacity in question only if we are able to identify other persons, and that, in turn, requires that something we can observe constitutes sufficient bases for such identification. But even if we accept Strawson's argument, it does not follow that our customary ways of forming beliefs about the conscious states of the persons so identified can be relied on to produce mostly true beliefs. More generally we can say that when recent Anglo-American philosophers have been engaging in producing arguments they call 'transcendental', they have been concerned with establishing the applicability of certain concepts, or the real existence of certain types of entities, or the possession of certain conceptual or judgmental capacities. They have not been concerned with the reliability or other epistemic properties of modes of belief formation. In a word, they have been doing metaphysics rather than epistemology. And so, despite initial impressions to the contrary, we can set their efforts to one side here.

This brings us to the end of our discussion of putatively a priori arguments for the reliability of SP. The chief morals to be drawn from this survey are the following. (1) Apparently a priori arguments frequently contain hidden empirical assumptions that rely on sense perception for their justification. (2) It is all too easy to (falsely) assume that by establishing various other things, such as the existence of the physical world, or the application of certain categories to the objects of experience, or the falsity of solipsism, or the impossibility of all perceptual beliefs being false, we can thereby show sense perception to be reliable. (3) Even if, contrary to my contentions, it were possible to give a conceptual backing for the principles of justification involved in SP, that would be only in an internalist sense of justification that carries no implications for reliability, which is what we are concerned with here.

Chapter 4

EMPIRICAL ARGUMENTS
FOR THE RELIABILITY OF SP

i. The Explanation of Sensory Experience

I now turn to a consideration of avowedly *empirical* arguments for the reliability of SP. First, a point made in chaper 2, section ii. Since we are looking for arguments that do not exhibit epistemic circularity, we are blocked off from using what is most usually termed 'empirical evidence', namely, what we learn about the environment from perception and anything based on that. This means that simple track record arguments and appeals to the predictive and control successes of SP are off limits.[1] Indeed, at first glance it seems impossible to mount any a posteriori argument for the reliability of SP without relying on its outputs to do so. For where else would we get the empirical premises we need for such an argument? There is at least one other source of knowledge, however that deserves to be called 'empirical', namely, introspection, our awareness of our own conscious states. This might be used to provide the empirical premises of an a posteriori argument, even a track record argument. If we could find some way to reason from facts about our sensory experience to external facts putatively perceived via sense experiences, without the conclusion depending on an appeal to our perception of any external facts, that might enable us to give a cogent noncircular argument for the reliability of SP. For it would enable us to show, with no empirical

[1]At least appeals to predictive success as these are generally carried out, in terms of the success of SP in predicting the course of events in the physical world. In the next section we shall see that a different kind of prediction, prediction of sensory experience, can be appealed to without epistemic circularity.

premises except those established by introspection, that, by and large, the facts we suppose ourselves to learn about by sense perception really do obtain. But how are we to infer external, perceivable facts from facts about one's sense experience? As has been repeatedly pointed out since the time of Hume, we can't proceed by appealing to sensory experience-external fact correlations that have been established inductively. For to perform such inductions we would have to have established by experience in a number of particular cases that, for example, when one has a sensory experience of type φ, there is a red object before one. Since we can experientially ascertain the latter part of this conjunction only by relying on sense perception, epistemic circularity creeps back in. However, there is another way. When high-level scientific theories are established (confirmed, tested, supported . . .) it is not by inductive correlations of the sort just mentioned, and for reasons analogous to those just appealed to. Consider a theory of chemical compounds in terms of constituent molecular structure. We cannot establish by induction that whenever we have something with the surface properties of salt, the molecular structure is thus-and-so; for apart from this or some other equally high level theory, we have no way of knowing what kind of molecular structure we have there. We have no cognitive access to molecular structure that is independent of a theory, just as in the present case we have no cognitive access to the putatively perceived facts independent of reliance on sense perception. But this difficulty by no means leads scientists to throw up their hands in despair. On the contrary, they have recourse to a form of argument called 'hypothetico-deductive', or 'argument to the best explanation'. Rather than seeking some inductive support for the theory, they try to show that it constitutes the best explanation of the empirical facts in question, in this case, inductively established lower-level lawlike generalizations. Similarly, it has often been suggested that the physical world as we normally suppose it to be provides the best explanation for our sensory experience or for certain introspectively ascertainable facts concerning our sensory experience. It is this project that we will be examining in this section.

All along I have warned the reader not to conflate the question of the reliability of SP with the question of the 'existence of the external world', and not to suppose that a successful argument for the latter would establish the reliability of SP. Those warnings are still in force. Now the argument we are about to consider is usually presented as an argument for the existence of the physical world. Nevertheless, in the course of developing the argument I will shape it in such a way that its conclusion does clearly imply that SP is reliable.

Before considering specific versions of the argument, it will be useful to remind ourselves how difficult it is to avoid epistemic circularity in the enterprise. Consider John Locke's famous discussion in the *Essay Concerning Human Understanding* (1975), which runs in part as follows.

But besides the assurance we have from our Senses themselves, that they do not err in the Information they give us, of the Existence of Things without us, when they are affected by Them, we are farther confirmed in this assurance, by other concurrent Reasons.

First, 'Tis plain, those Perceptions are produced in us by exteriour Causes affecting our Senses: Because *those that want the Organs of any Sense, never can have the* Ideas *belonging to that Sense* produced in their Minds. . . . The Organs themselves, 'tis plain, do not produce them: for then the Eyes of a Man in the dark, would produce Colours, and his Nose smell Roses in the Winter: but we see no body gets the relish of a Pine-apple, till he goes to the Indies, where it is, and tastes it. (P. 632)

Fourthly, Our *senses,* in many cases bear *witness* to the Truth of each other's report, concerning the Existence of sensible Things without us. . . . Thus I see, whilst I write this, I can change the Appearance of the Paper; and by designing the Letters, tell before-hand what new *Idea* it shall exhibit the very next moment, barely by drawing my Pen over it: which will neither appear (let me fancy as much as I will) if my Hand stands still; or though I move my Pen, if my Eyes be shut: Nor when those Characters are once made on the Paper, can I chuse afterwards but see them as they are; that is, have the Ideas of such Letters as I have made. Whence it is manifest, that they are not barely the Sport and Play of my own Imagination, when I find, that the Characters, that were made at the pleasure of my own Thoughts, do not obey them; nor yet cease to be, whenever I shall fancy it, but continue to affect my Senses constantly and regularly, according to the Figures I made them. To which if we will add, that the sight of those shall, from another Man, draw such Sounds, as I before-hand design they shall stand for, there will be little reason left to doubt, that those Words, I write, do really exist without me, when they cause a long series of regular Sounds to affect my Ears, which could not be the effect of my Imagination, nor could my Memory retain them in that order. (Pp. 633–34)

The first of these concurrent reasons is the most blatantly circular. How does Locke know that those without the use of their eyes never have visual ideas, except by relying on what he has learned through perception, including what other people say? Indeed, so sweeping a generalization could not be based solely on what he has observed, but requires crediting the testimony of others, which itself is known about only through perception. For that matter, how does he know that anyone has, or lacks, eyes? The same point applies to such claims as that the eyes of a man in the dark do not produce colors, and that nobody gets the taste of a pineapple without putting that fruit in his mouth. The circularity in the fourth reason is a bit less obvious, but just as unquestionably there. To be sure, one could, without circularity, appeal to correlations between, for example, visual and tactual experiences in arguing for a common physical cause; but Locke's presentation goes well beyond that. He appeals to the fact that his visual experience will be different depending on whether *his hand moves his pen* and on whether *his eyes are shut*. But once more he can't tell whether his hand is moving his pen or whether his eyes are shut without relying on sense perception (including internal perceptions of bodily goings-on). Again, in appealing to what his inscriptions will elicit from another person, he is relying on sense perception to ascertain both that there are those inscriptions and that *another person* reacts to them in a certain way.

Now for arguments that I will take more seriously. Attempts to show that the course of our sensory experience is best explained in terms of physical causes differ both in how they construe the explanandum and in how they construe the explanans. I shall begin with the former.

First, note that we can't get any mileage at all out of considering particular bits of experience one at a time. If I take a particular visual experience and consider what is responsible for my having that experience, without taking into account how it fits into some overall pattern, I will have no basis for choosing between different possible explanations, apart from whatever differential credibility they have from other sources. So far as throwing light on the fact that this experience occurs, so far as *the connection between explanans and explanandum* is concerned, one explanation is as good as another when the explanandum is this impoverished. If I can make the effort of will necessary to leave out of account everything else I think I know about the course of my sensory experience and whatever I base on that, I will be completely at a loss in choosing between the usual physical-physiological-psychological explanation, the Cartesian demon explanation, the self-generation-by-unconscious-psychological-processes explanation, and others. What possible basis could there be for such a choice? I have a reason for picking one explanation of X rather than another after some lawlike generalizations have been established connecting states of affairs like X with others. If I have succeeded in correlating combustion with other observable phenomena, for example, heat and the need for an air supply, I have something to go on in determining what underlying mechanisms are responsible for combustion. But without such data we have no idea where to turn. Thus we are well advised to choose a richer explanandum, such as recurrent patterns we find in our sense experience.

On the side of the explanans it also turns out that the simplest choices fail to provide reasons for taking SP to be reliable. A minimal explanatory gesture in the direction of an external cause would be that some (relatively) permanent and stable entity or entities is causally responsible for our sensory experience. But even if that were true, it would fail to show that these entities are as they perceptually seem to us. That explanation is compatible with the cause being God, a demon, Kantian things-in-themselves (with the ban on causal efficacy lifted), or Leibnizian or Whiteheadian centers of psychic force. The explanation could be enriched with the stipulation that the cause(s) are physical in character, but that would still not show that sense perception gives us the correct story about those causes. It could be that our sensory experience is due to the causal impact of physical things on us, but not in such a way as to yield generally correct beliefs about those causes. This could be either because the same external state of affairs does not, even usually, produce the same kind of sensory experience (because of the preponderant influence of factors internal to the subject), or because of peculiarities in the perceptual belief-producing mechanism. In the latter case, the experience itself could be as sensitive an indication as you like of the nature of its causes, but the belief-producing mechanism(s) fails to exploit this fact sufficiently.

Thus we must enrich the explanans still further. Rather than a bare supposition that sensory experience is caused by something(s) physical, we must build in the supposition that sensory experience is caused in pretty much the way contemporary science thinks it to be caused. That will give us a much richer and more detailed explanatory hypothesis, one that is open to continual development and modification but that is clear enough in its general outlines.

But will even this explanation imply that sense perception is generally reliable, or give strong support to that supposition? At least this much is clear: If our sensory experience is caused in the way we currently suppose it to be, then so long as we are in normal environments,[2] differences in our experience will consistently reflect differences in the parts of the causal chain concerning which perceptual beliefs are normally formed. That is, when two visual experiences differ in respect G, their causes, or those parts of the total causes that are perceived and concerning which beliefs stemming from those experiences are typically formed, will differ in respect H, and this correlation will hold generally. Thus if one visual experience is of a blue expanse and another is of a red expanse, then the objects thereby perceived will differ in a way that is typical of objects that, in normal settings, cause a visual experience of a blue expanse and of a red expanse, respectively. And so if sense experiences are caused as current science supposes, they serve, in normal environments, as reliable indicators of certain features of certain of their causes.

But will this insure that perceptual beliefs are generally correct? Well, that depends on a couple of further points. First, it must be the case that perceptual beliefs are formed in such a way as to exploit the differential sensitivity of sense experience we have just described. In other words, our normal belief-forming mechanisms will have to be appropriately sensitive to differences in the sensory experiences they take as input. If that is so, can we infer that perceptual beliefs are by-and-large true, at least when formed in normal environments? That depends on how those beliefs are to be understood. Consider color attributions. Just what am I claiming when I say, on the basis of seeing

---

[2]This qualification is needed because if the setting is abnormal enough in certain ways, the differences in sense experience will not reflect the kinds of differences in the perceived objects they consistently reflect in normal environments. Thus, if we are looking at things under unusual artificial light or at holographs, or, still more, if the experiences are being produced by direct brain stimulation, then differences in the experiences won't reflect the usual differences among putatively perceived objects.

X, that X is red? Am I merely claiming that X has some feature the intrinsic nature of which I am leaving open, a feature that is typically found in seen objects when they are involved, in the usual manner, in causing a visual experience in which something appears to be red? Or am I, more boldly, attributing a certain intrinsic quality to X, the very quality exhibited in the way in which X visually appears to me? If the former, then it would seen that the assumptions we have made will guarantee the by-and-large correctness of color attributions, when made in normal environments.[3] But if the latter, no such implication holds. In fact, it has been widely held since the seventeenth century that color attributions, so understood, are invariably false, even if color experiences do consistently reflect physical differences in the object. Lacking the time to go properly into the thorny issue of the status of color and other "secondary qualities" of physical objects, I shall simply rule that perceptual color judgments are to be taken in the former, more modest fashion.

The considerations of the previous paragraph can be generalized to other "secondary qualities" like sounds and their variations (pitch, loudness, timbre), "feels" like degrees of heat and cold, odors, and tastes. In all these cases it is plausible to adopt the more modest interpretation according to which attributions to perceived objects are to be understood simply as attributions of unspecified properties that are regularly reflected in sensory appearances with the qualitative distinctiveness in question. But when we come to primary qualities, attributions of intrinsic qualities cannot be so easily brushed aside. It is not at all implausible to suppose that when I perceptually judge an object to be round or triangular, or when I judge A to be closer to B than A is to C, I do mean to be saying something about how it is with these physical objects in themselves, and not just that they have *some* property, or are related in *some* way, that is reflected in sensory appearances. But, fortunately, there are not the same reasons here as in the case of secondary-quality attributions to doubt that such judgments are often true. There are not the same reasons for denying that physical objects do have shapes and relative spatial position of

---

[3]For an opposing view, see Hardin (1988). But Hardin acknowledges that attributions of color to physical objects can serve, in most typical situations, as a rough and ready practical guide to our dealings with objects.

the same general sort they sensorily appear to have, even though the details in particular cases may vary.

Let's draw together the threads of this discussion. If our sense experiences are normally formed by the impact of stimulation from the physical environment, as that environment and that impact are currently conceived by scientifically enlightened common sense, then differences in our sense experiences will, in normal situations, reflect in some general, lawlike fashion differences in the environmental objects and situations putatively perceived therein. And that will, in turn, imply that our perceptual beliefs, on a natural interpretation thereof, will be generally correct, provided two further assumptions hold.

(1) Our perceptual belief forming mechanisms are themselves sufficiently and appropriately sensitive to differences in their experiential input.

(2) We are usually in normal situations, situations in which nothing is throwing off the tendency of the total apparatus to produce mostly true perceptual beliefs. For even if things are with perception and perceptual belief formation as we ordinarily suppose, we could hardly expect people to have mostly true perceptual beliefs if they were mostly confronted with extremely clever look-alikes, or if their sensory experiences were frequently enough produced by direct brain stimulation in a laboratory.

Thus, in order to insure that our explanatory hypothesis will imply the reliability of SP, we must build in these two assumptions as well. To be sure, it could hardly be part of a scientific account of perception and perceptual cognition that sense experience is usually produced in normal environments. It is no part of science to determine where people spend most of their time. The physics, physiology, and psychology of perception could be true or adequate just to the extent that they in fact are, even if the majority of percipients spent most of their lives in subjection to neurophysiological experiments in which their experiences are artificially produced by direct brain stimulation. But what does or does not belong to science is of no import for our present concerns. We are not trading on the results of science or appealing to the authority of science in putting forward the explanatory claim in question. On the contrary, the terms of our project forbid us from borrowing any support from scientific results, on pain of epistemic circularity. We are merely using science as a source of suggestions, which will then have to be evaluated without any supposition that science itself is well founded. Hence, we are free to insert anything we like into the explanatory hypothesis under consideration. That does not mean, of course, that we are free to make any judgment we like as to the adequacy of the explanation or as to its standing relative to its competitors. Anything we insert into the hypothesis will have to pull its explanatory weight. If it adds nothing of explanatory value, the hypothesis will suffer by comparison with a leaner competitor. With all that in mind, let's take our explanatory hypothesis to include the following three components.

(1) Sense experience is produced in the way currently supposed by scientifically enlightened common sense.

(2) Our belief-forming mechanisms are properly sensitive to differences in their experiential inputs.

(3) Sense experience is usually produced in normal environments.

Before getting down to business we must attend to another complication. So far, we have been concentrating on the simplest kind of perceptual beliefs, those that involve the attribution of simple sensible qualities, both primary and secondary. But perceptual beliefs commonly involve the identification of *things,* both as to their kind and as to their individual identity; and they also commonly involve attributing more complex properties, like elegance, provenance (manufactured by Ferrari or composed by Handel), and material composition (pure leather). We have the capacity to determine by looking or listening that something is a house, is Susie's house, is a Beethoven string quartet, is made of plastic, and so on. Moreover, in many cases we arrive at this belief without any conscious inferential process. Here there is an even more tortuous route from the standard explanation of sensory experience to the reliability of the belief-forming process. It is clearly possible that sensory experience should be produced in the way we normally suppose and yet people might not be very good at recognizing their friends or their friends' houses, at identifying musical compositions, or at determining the material of which something is made.

Thus, we must decide whether an explanatory hypothesis adequate for our purposes need only imply that sense perception is a reliable source of attributions of sensible qualities, or whether it must also imply that sense perception is a reliable source of more complex perceptual beliefs. In other words, is the practice the reliability of which is in question solely a practice of forming sensible-property attributions on the basis of sense experience, or is it, more broadly construed, the practice of forming any perceptual beliefs about the external environment (or perhaps any perceptual beliefs not based, even in part, on inference)? I could make either choice. If I chose the more restricted practice, the explanatory hypothesis could be less ambitious. But, on the other hand, what is left out of that specification is a good part of what interests us most. We are concerned not only with whether we can reliably attribute colors and shapes, feels and sounds, to things, but also whether we can identify the things we encounter in ways that are of practical importance—as this or that person, as a golden retriever or a pit bull, as a TV set or a machine gun. Hence, I will opt for relevance and construe our problem as concerning the more inclusive practice of perceptual belief formation. This means that the explanatory hypothesis must be correspondingly enriched. It will have to include not only the supposition that our usual perceptual belief-forming mechanisms are such as to reflect the way sensory experience is an indication of the presence or absence of simple sensory qualities, but also the much more complex supposition that these mechanisms are such as to go from appropriate inputs to generally correct beliefs about the identity and character of the kinds of objects about which perceptual beliefs are formed. To spell out the detailed character of such mechanisms would be a forbidding and perhaps impossible task. Fortunately, this is not necessary for our purposes. We can continue to construe SP as made up of the perceptual belief-forming mechanisms that we generally or normally use, and consider the question of whether the assemblage of such mechanisms produces, by and large, true beliefs.

ii. The Explanation of Patterns in Sense Experience

We are now in a position to begin our assessment of the explanation under consideration. For easy reference, call it the 'standard explanation'. Contrary to what one might naturally suppose, it cannot assume the form typical of theoretical explanations in science, in which a body of lower level, inductively established lawlike generalizations are explained by hypotheses concerning the underlying structure and/or mechanisms that are responsible for these regularities. Consider, as a modest example, the vast body of lawlike regularities in the chemical interaction of substances, where the substances are characterized in terms of empirically accessible surface properties, like color and weight, or in terms of common-sense natural kinds, like milk and salt, and where the interactions and their products are also described in observational or near-observational terms. These interactions include combustion and other forms of oxidation, the reduction of metallic ores, corrosion, and so on. A theory is then developed that identifies elements by atomic weight and the like, and describes compounds in terms of molecular composition. None of this latter is directly accessible to observation, even with the help of instruments (unless the instrumental readings are informed by the high-level theory). Acceptance of the theory is justified by the fact that it makes possible a systematic unification and explanation of a vast body of otherwise heterogeneous and unrelated empirical regularities.

The present explanatory enterprise cannot be construed in these terms just because there are no genuine lawlike regularities at the purely phenomenal level.[4] We cannot ascertain any regularities that we have reason to think will hold up under any and all circumstances, so long as we stick strictly to sensory experience. The first point to note here is that we cannot identify even rough regularities so long as we leave out of account the behavior of the subject. I am looking at my desk in my study with a cork bulletin board behind it on the wall. What will follow this visual display? That all depends, obviously, on

---

[4]Robert Audi has suggested to me that it might be a phenomenal law that a very loud sound is followed by the sensation of "ringing in the ears". But even if there are a few cases like this, it doesn't suffice to invalidate the general point being made here.

what I do. If I remain motionless with eyes open I will continue to be presented with an almost identical visual field, at least until some sensory satiation point is reached. If I turn my head in one direction, something different will follow; if I turn my head in another direction or if I move into another room, still different visual experiences will ensue. It is hopeless to try to find regularities in changes of sensory experience without taking into account the movements of the subject. But to take those into account we will have to abandon purely phenomenal description, will we not? Well, insofar as kinaesthetic and other sense experiences are an accurate reflection of bodily movement, something may be done. By factoring these phenomenal shadows of the subject's locomotion into the equation, we may be able to sketch out rough regularities. Thus, my visual experience as if looking at the wall behind my desk plus certain characteristic kinaesthetic sensations will be regularly followed by a visual experience as if seeing the door of my study with a full bookcase to its right. And so on.

But no such regularities have any title to lawlikeness. They are eminently subject to exception even when the phenomenal antecedents are as specified. If the bookcase to the right of the door has been moved, the visual consequent mentioned in the previous paragraph will not be forthcoming. If my neighbor across the street cuts down his trees, the visual scene to the right of my desk will have changed. When I open my eyes in the morning, I am confronted with a typical visual experience of my bedroom. But I have lived in other houses, where what I experience under those conditions is quite different. All such regularities—constancies as well as successions—are at the mercy of transactions in the external environment that are not registered in my experience. Moreover there is no pattern to these exceptions that would give us a handle on taking account of them in some principled fashion.⁵

It may be thought that we will achieve lawlikeness if we turn from the "local" regularities just exemplified to the phenomenal shadows of physical lawlike connections. A visual experience as of putting a lighted match to some newspaper will be followed by a visual experience as of the newspaper catching fire. A tactile experience as of

⁵I owe this last point to Robert Audi.

grasping a door handle and turning will be followed by a visual experience as of a door opening. A tactile experience as of a gust of wind plus a visual experience as of leaves on the ground in the vicinity will be followed by a visual experience as of leaves moving. But here, as in the earlier examples, there is not enough in the phenomenal shadows to effect lawlike connections. The newspaper may be flame resistant in ways not detectable by just looking, or the apparent flame of the match may be an optical illusion. Doors notoriously sometimes fail to open when we turn the knob and pull. The leaves may be firmly stuck to the ground or the wind may not be strong enough to move them. Perhaps we could get rid of all these exceptions by suitable enrichments of the phenomenal antecedents. But further exceptions are always waiting in the wings.[6]

Perhaps the closest we can come to genuine phenomenal laws concern variations in visually sensed shape and other spatial properties with variations in relative position and distance of observer and object. Suppose we really can take care of observer movement in terms of kinaesthetic and other sensations. Then it is a plausible candidate for a lawlike regularity that as the observer moves away from a position directly in front of an object that presents a round appearance, that object will appear more or less elliptical. (To be sure, this is complicated by constancy mechanisms, and there is persistent controversy over whether there is an increasing *experienced* ellipticality that is masked by belief and other nonperceptual cognitive factors.) But even if there are a few phenomenal laws of this sort, they are too few and too isolated to provide the explanandum base for an explanatory theory rich enough to bring in what it takes to support the reliability of SP.

Even if there are no, or few, lawlike regularities within the phenomenal realm, that is not necessarily the end of the explanatory enterprise. It would be a narrow, dogmatic scientism that would allow as explanations of sensory experience only what conforms to the pat-

---

[6]It may be pointed out that even with physical descriptions, practically any laws we are able to work with hold only within a "closed system", closed against any effective influences not taken account of in the law. But the point is that with physical variables we are able to specify connections that are so close, for all practical purposes, to holding no matter what, that it is at least close enough to the truth to think of ourselves, much of the time, as working with a closed system. Such is decidedly not the case with the kinds of putative phenomenal laws we have been illustrating.

tern of theoretical explanation in the most advanced sciences, where we have genuinely lawlike empirical generalizations with which to work. We often seek to explain rough regularities in everyday life, and such explanations often genuinely throw light on the matter. We seek to explain the fact that Jim is often so moody on weekends, we ask why the furnace tends to break down when the weather is coldest, why people catch more colds in winter than in summer, and so on. Why shouldn't we seek an explanation of the rough, and even local, regularities we find in our sensory experience? And perhaps the best explanation we can find is one that will imply the reliability of SP.

This brings us to the question of whether the standard explanation of patterns in sensory experience is superior to its competitors. Remember that we are ultimately interested in an explanatory hypothesis that not only involves the standard suppositions about the production of sense experience but also contains the assumption of a standard mode of perceptual belief formation, as well as an assumption of the by-and-large normality of situations in which perceptual beliefs are formed. However, we would be best advised to focus on the production of sensory experience. Since virtually all the discussions in the literature are confined to the explanation of sense experience, this will enable us to make contact with that literature. Since this part of the job is fundamental to the rest, if it fails here, as we shall see it does, that will effectively subvert the whole enterprise.

At the risk of belaboring the obvious, I will make explicit just how it is that the standard explanation of sense experience bids fair to throw light on the kinds of patterns cited earlier. First, consider the constancies and regularities of succession in my visual experience when certain kinds of kinaesthetic sensations do (or do not) occur (i.e., though we can't explicitly say this, when I remain at rest or move in certain ways). When I have a visual experience as if seeing the desk in my study, this will remain constant, then be followed by certain other visual experiences, and reappear, depending on the kinaesthetic sensations that occur. These patterns are naturally explained by the supposition that the experiences are produced in the standard way, thereby revealing an external environment that is relatively stable and structured in such a way as to yield visual experiences of the sorts specified depending on what visible objects are in the line of sight and close enough when I turn my open eyes in one or another

direction. Again, the spread of fire and other causal rough regularities in experience mentioned earlier receive a ready explanation
if we suppose these experiences to be caused by environmental hap-
penings in such a way as to reflect the character and causal interactions of the physical realities involved. Finally, consider the continuous deformation of visually sensed shape as we get the sensory
correlates of movements away from looking straight on to an object,
and the fact of increasing apparent size as we, so it seems sensorily,
move closer and closer to an object, with a typical set of tactual experiences becoming available as the object looms largest in the visual field.
All such patterns are naturally understood in terms of the experiences being produced in the standard way by external objects that are
spatially disposed in a certain way at each moment vis-à-vis the moving observer. I take it that there is no question but that the standard
theory has explanatory force with respect to these explananda.

But the question remains as to whether it is a better explanation
than its competitors. Just what competitors are there? Discussions of
the issue have been almost hypnotically fastened on the Cartesian
demon or the Berkeleyan God, according to theological taste. It has
been argued, as we shall see in a moment, that the standard explanation better satisfies certain generally accepted constraints on explanation than do these alternatives. But even if that is so, we are not home
free unless other live alternatives have been excluded. What others
deserve consideration? The phenomenalist view that patterns of experience are ultimate, requiring no explanation in terms of anything
else, is often mentioned; but I will summarily dismiss this null hypothesis, on the grounds that it takes us outside the explanation
game.[7]

Another large class of explanations that is rarely given so much as
a nod comprises those that invoke physical causes with quite different
properties from those we believe ourselves to perceive; there will, of
course, have to be suitably different laws to make it possible for these
bodies and forces to produce our sense experience. The range of such
explanations is limited only by our ingenuity. Let's unreservedly ad-

---

[7]Actually the situation here is more complicated, since the usual phenomenalist account
is in terms of patterns of actual *and possible* experiences; and the appeal to possible experience
raises problems of its own. These problems will, however, do nothing to nullify the reason for
which I dismissed such accounts.

mit that no such explanations possess any empirical or scientific support or even initial plausibility; but that is nothing to the point. Remember that, by the terms of the enterprise, we cannot call upon anything we have learned from sense perception and reasoning therefrom; that applies to the winnowing out of candidates for explanation as well as to arguments in their support. It may be that some explanations of this class will imply that SP is as reliable for beliefs about secondary sensory qualities as the standard explanation does. For if such beliefs are, as we have been supposing, to the effect that the perceived object has some property or other, identified only as the property the possession of which by an object disposes it to produce experiences with a certain phenomenal quality, then such beliefs could be correct even if the properties in question were intrinsically very different from what they are supposed by contemporary science to be. But even alternative physical explanations for which this is true could be such as to imply that other perceptual beliefs yielded by SP are more often false than true. And there could be still other alternative physical explanations that would not make provision for enough regularity in the physical property-phenomenal quality correlations to imply reliability of secondary-quality attributions even on this modest interpretation. And so we will have to recognize alternative physical explanations as an indefinitely large set of possible competitors.

There are other possible explanations that are like these, and unlike appeals to deities or demons, in that they represent sense experience as due to the causal impact on the subject of a variegated environing world of finite substances on roughly the same scale as physical substances, but that differ in that they are not material or physical in character. The most familiar view of this sort is Leibnizian or Whiteheadian panpsychism. According to both there is no such thing as (completely) dead matter. Every existing entity engages in *perception* (Leibniz) or *feelings* (Whitehead), though these rise to a conscious level only in such high grade existents as human minds. What perception inaccurately presents to us as a field of inert, lifeless things (except for the plants and animals therein) is really a continuum of enduring perceiving substances (Leibniz) or a four-dimensional continuum of "occasions" of experience (Whitehead). Our interaction with these beings gives rise to our sense perceptions, which, taken uncritically, lead us to form false beliefs about the character of the entities we perceive.[8] Here too we have a class of explanations that are genuine alternatives to the standard explanation and that do not imply that SP is reliable, and in fact imply that SP is *unreliable*.

To this it may be objected that, according to Leibniz, our perceptual beliefs, however false on a strict reckoning, do serve us well practically, since the envisaged spatial field in which physical substances display themselves is, in Leibniz's phrase, a "phenomenon bene fundatum," a well founded phenomenon, a faithful mirror of the real field in which the real psychic substances are interrelated. And so long as perceptual beliefs serve us well as guides to action, we cannot properly take SP to be unreliable. But reliability requires a preponderance of *true* beliefs, not just practically useful beliefs; and that is precisely what we do not have on the panpsychist explanation. It is true that one reason we value true beliefs is that they provide us effective guidance in our interactions with the environment. And if a set of false beliefs should do equally well in this regard, a contrasting set of true beliefs would not enjoy that advantage. Nevertheless, there is still an intrinsic value of believing what is true rather than what is false.

A more conclusive reason for denying that the Leibnizian alternative implies the unreliability of SP is that in explaining the notion of reliability near the beginning of this essay I said that it would not require beliefs that were perfectly true, but only beliefs that made a sufficient approximation to the truth. And it may well be argued that in Leibniz's view ordinary perceptual beliefs satisfy that condition. Thus, I will break off the attempt to claim that Leibnizian panpsychism implies the unreliability of SP, and instead point out that we can envisage panpsychist explanations that lack the codicil that perceptual beliefs provide effective guidance in our interactions with the environment. If it is replied that such explanations could be ruled out on the grounds that our perceptual beliefs do obviously provide such guidance, the answer is that this bit of information is not available to us in this discussion, since our acceptance of it rests on observational evidence. If we were to appeal to it, we would once again be en-

---

[8]This talk of "interaction" does not represent Leibniz with strict accuracy; he denies that substances can affect each other causally, though because of the pre-established harmony, their successive states are correlated as they would be if they did causally affect each other.

meshed in epistemic circularity. Thus I will count (some) panpsychist explanations as another competitor to the standard explanation.

Another alternative that does not get the innings it deserves is the self-generation hypothesis. It is conceivable that my sense experiences are wholly generated by myself. To be sure, I am not conscious of doing so; but then I do many things I am not conscious of doing, metabolic processes, for example. Obviously, this explanation would not imply that SP is reliable, and if we exclude pre-established harmony, it will imply that SP is unreliable.

Let's say then we are working with the following alternatives to the standard explanation.

(1) Cartesian demon explanation.
(2) Berkleyan God explanation.
(3) Alternative physical explanations.
(4) Panpsychist explanation.
(5) Self-generation explanation.

(1) and (2) differ in that whereas the Cartesian demon is motivated to deceive us by directly producing our sensory experiences in such a pattern as to lead us to erroneously suppose ourselves to be perceiving an environing physical world, the Berkeleyan God has no such deceptive purposes. It is just that we are led by a false metaphysics of our own construction to put a false construal on what we perceive. Of course, given the pervasiveness of this construal, Berkeley is faced with the Cartesian theological question as to how we can exculpate God from a charge of deceptiveness. Berkeley would reply that the error is not very pervasive, being confined to a few misguided theorists; the overwhelming mass of humanity understand perceptual judgments in such a way that they are mostly true on a Berkeleyan interpretation. I shall assume that Berkeley is mistaken about this, and that if his explanation of sense experience is correct, SP is most decidedly not reliable. Hence I do not take the differences between (1) and (2) to be especially relevant here; the two explanations will be contrasted with the standard explanation in the same way.

The reader may be surprised that I did not include the currently popular "brain in a vat" hypothesis, according to which all my sensory experiences are produced by direct brain stimulation, I myself being an isolated brain kept alive in a nutrient solution. The reason is that such a hypothesis inevitably raises the question of the status of the perceptions of the scientists and technicians who are thus manipulating me. Presumably, in order to make the hypothesis readily intelligible and not intolerably implausible, they will have to be credited with normal perception. But then the hypothesis does not imply that SP is *generally* unreliable. Even if other sentient subjects are in the same unhappy situation as I, there would have to be some normal percipients to keep the thing going. Thus, for deliberate deception I will restrict myself to the good old Cartesian demon, who can merrily deceive all humankind evenhandedly.

iii. Attempts to Support the Standard Explanation

Contemporary discussions of this problem typically assume that all these competitors are empirically equivalent in that they all explain the same range of experiential data. The task of choosing a winner then boils down to determining which candidate best satisfies other criteria for the comparative evaluation of explanations. Those most commonly cited are economy and simplicity. But it is not clear that the alternatives all score worse in these respects. The criteria in question are notoriously subject to various construals, but let's say that economy is a matter of the number of entities, or, better, the number of basic kinds of entities, to which the explanatory hypothesis commits us. On this score it would seem that (1), (2), and (5) win hands down over the standard theory and (3) and (4). Each of the former postulates only a single kind of explanatory factor, indeed only a single individual factor. Whereas the standard view, along with (3) and (4), is recklessly prodigal in the kinds of things it invokes: as many kinds of things as we currently recognize to inhabit the physical world. Simplicity is a more difficult notion to make precise. In default of a thorough discussion, let's say that it is a matter of the complexity of the connections between the ultimate explanatory factors and the terminal explanandum. On this reading (1), (2), and (4) are again clearly superior, or can easily be construed as such. Let's say that God or the demon or the subject herself produces sensory experiences directly. In the first two cases, a simple act of will suffices. In the third case, some unconscious analogue of an act of will can be deemed sufficient. But perhaps simplicity is something like the complexity of the whole story that has to be told in order to complete the explanation. On that understanding, it has been argued that the demon hypothesis is less simple than the standard one; for to bring it about that the hypothesis suffices to explain our experience, we must represent the demon as using the standard scheme as a model so as to achieve his deceptive aim to give us a pattern of experiences that will lead us to believe we are perceiving a physical world. But it is not at all necessary that a demon proceed in this way. Why couldn't he simply intend to produce in us experiences that exhibit the kinds of internal patterns we have already noted, along with giving us the penchant to go from such patterns to beliefs about physical things supposedly perceived in these experiences? The demon need not spell out in advance all, or even any, of the details of the particular physical world scheme we end up with, though of course it is conceivable that he should do so.[9] Thus the prospects for justifying the standard explanation by an appeal to economy or simplicity do not look promising.

Alan Goldman (1988, chap. 9, "The Inference to Physical Reality"), appeals to "explanatory depth" as a criterion for goodness of explanation. He initially compares the standard explanation with the view that phenomenal regularities are ultimate, and rightly points out that the former is deeper in that it answers questions (why does our experience exhibit just these regularities?) that the latter refuses on principle to raise (pp. 206–7), as well as explaining deviations from the regularities, not just the regularities themselves (p. 208). But, more to the present point, although he recognizes that alternatives like the ones we are considering also provide deeper explanations than phenomenalism (pp. 209–10), he objects to deceiver hypotheses like (2) on similar grounds.

> If God caused all the lawlike regularities in our experience without creating physical objects, then, as Descartes pointed out, he would be a

---

[9]It may well be that *we* cannot set out the demon's program without using our standard physical world scheme as a model for the patterns of sense experience he aims to produce in us. But there is no reason to suppose the demon to be restricted in this way.

deceiver. At least his motives would be far from clear, as unclear as the mechanisms by which he worked this enormous deception. The same obscurities as to motives and mechanisms, that is, as to any attempt to carry the purported explanations to deeper levels and make them more explicit and detailed, infect the superbeings of the skeptic's imagination. Not only do we not know why or how they produce the sequences of appearances for us that they do; we also have no idea why they make it seem to us as if there are physical objects. The latter question, natural here, does not arise given the initial explanation of appearances that we all naturally accept, and accepting the explanation as true leads to deeper and more satisfying theories to explain physical interactions. Certainly whatever psychology and programming apparatus one could dream up for the superbeings could not be as rich or testable as the theories of physical science and neurophysiology. (P. 212)

There seem to be a number of points jumbled together here. First, "we do not know why or how they produce the sequences of appearances for us that they do". If this means that we do not have independent information as to this why and how, that is irrelevant. As we have noted before, we can't make use of such independent information with respect to the standard theory either, or with respect to any other alternative, without epistemic circularity. When we are at a level this basic, we may concoct any explanatory hypothesis we choose, *without antecedent evidence for its truth,* and then let its relative explanatory efficacy decide whether it is to be accepted. But perhaps the point is that while the standard theory is enormously rich and complex, the alternatives, as thus far developed, consist merely in bare suggestions as to how an explanation might be developed in detail. Thus, in comparing actually existing theories, the standard one is miles ahead on richness and detail. This is an important point. But what bearing does it have on the question of which explanation is the true one? Obviously, the reason the other alternatives have not been significantly developed is that we all accept the standard explanation and are quite satisfied with it. No one is motivated to carry out the enormous labor that would be involved in developing one of the alternatives to anything like the same extent. But to build this superiority of the standard explanation into an argument for its truth, we would have to be justified in supposing that (a) none of the other alternatives could be developed into something equally rich and detailed, and (b) of two explanations the richer and more detailed is more likely to be true, ceteris paribus.

(b) may be defensible. This is one form of the general question as to whether these "good-making" properties of explanations should be thought of as indications of *truth,* or of other desiderata: fruitfulness, ease of operation, intellectual satisfaction, or whatever. Perhaps this question should be answered differently for different criteria. The truth conducivity of simplicity and economy is often questioned. Why should we suppose that reality is maximally simple or maximally economical? We can readily understand the appeal of these criteria without supposing them to be indicative of truth. A simpler theory is easier to grasp and easier to work with; a more economical theory presents us with fewer complexities to keep track of. One might say something similar about *depth* in the sense in which I first reported Alan Goldman as using it, a sense in which it amounts roughly to pushing explanation as far as possible. (And who is to say how far is possible?) Again, we can understand the appeal of this without supposing that it has anything to do with truth. If explanation is our game, we want to keep doing it as long as possible, or we are out of business. But why suppose that an explanation in terms of a further level, realm, or domain is more likely to be true, ceteris paribus, than one that stays modestly at home? There is more to be said for the desideratum currently under consideration, also termed 'depth' by Goldman, but better termed 'detail'. This is, in a way, the opposite of simplicity. On one natural understanding of 'simplicity', detail and simplicity are contraries; the simpler the hypothesis the less detail and vice versa. Before considering whether *detail* is truth indicative, we had better render our understanding of it more detailed. Are we speaking of detail in the explanans or detail in the connection? If it is only the former, then if we compare two hypotheses that are unequally detailed but have equal explanatory power, I think we must prefer the less detailed one; it gives us more bang for the buck. To have a clear ground for preferring more detail, I think it will have to be construed in terms of the explanatory connection. H_1 is more detailed than H_2, and this additional detail enables us to explain more detail in the field of the explanandum. Here we have a clear superiority and one that arguably increases the probability of truth, ceteris paribus. But note that this superiority is simply a consequence of the greater explanatory power. It is not the detail as such, but the resultant explanatory power that makes it a better explanation. That may be what Goldman was thinking: that the standard theory can throw more light on why we have patterns of each of a number of different sorts, and can differentially explain each by specifying the causal antecedents peculiar to it; while the demon hypothesis can only do a bit of hand waving with its talk of an intention to deceive.

Is there really this difference between the standard explanation and alternative (1)? ((2) can be treated in the same way). So far as I can see, that all depends on whether we require (1) and (2) to be developed without reference to the details of the standard explanation. Without any such reference, then (1) and (2), in any actually existing form, simply amount to the very unspecific suggestion that the demon (God) directly produces our sensory experience in such a way as to lead us to form the belief system we have. That's not much explanatory detail. But suppose the explanation is spelled out in the following way. We take over the standard explanation, lock, stock, and barrel. We then think of the demon (God) as directly producing our experiences on that model; that is, the demon produces our experiences just as they would have been produced if . . . (what follows is all the detail of the standard explanation). Spelled out in this way it loses its superiority on the grounds of simplicity, for it is now even more complex than its standard rival. But it doesn't suffer from lack of detail. By the terms of the enterprise, it includes all the explanatory detail of the standard scheme.

Goldman may try to counter this by downplaying (1) on the grounds of derivativeness, lack of originality. It doesn't really amount to the working out of a genuine alternative, one that stands on its own feet as far as the details of the explanation are concerned. It comes to no more than riding piggyback on the standard explanation—giving it a new twist as a whole. This is certainly an accurate picture of the situation I have just sketched, but it remains to be seen what bearing it has on the comparative evaluation of the theories. It has a clear bearing on the comparative evaluation of the *theorists*. One who put forward a demon explanation in this form could hardly lay claim to an intellectual achievement comparable to the aggregate of those who have developed the standard explanation. But that is neither here nor there, so far as the merits of the explanations are concerned. One who makes a small but crucial modification of standard quantum theory may have demonstrated much less scientific genius, ingenuity, or creativity, than the pioneers of that discipline; but for all that, his version of the theory may be clearly superior. And so it is here. Even if (1) is just the standard explanation with a twist, it may still be that the twist makes it a superior explanation. To deny this on the grounds of inferior originality is like badmouthing Japanese electronic products on the grounds that they got the basic science and technology from us in the first place. Even if that is true, it can also be true that their utilization of this technology is such as to issue in superior products.

The remarks of the preceding paragraph apply to (5) as well. The self-generation explanation can appropriate the standard explanation in just the same way. It can take the form of saying that each subject generates his/her own sense experience in just the ways it would be generated on the standard scheme. It is not clear whether (4) could be treated in the same way. Presumably any interesting panpsychist theory would have to say something about the way in which the distinctively psychic character of the environing perceived world enters into the production of our sensory experience. And as for (3), the piggyback possibility would seem to be closed. This is, as we might say, on the same level as the standard explanation, differing in the nature and operation of the physical factors it postulates. Hence these will have to be specified in detail before we have a genuine alternative explanation of this sort.

Thus with respect to explanatory detail, our alternatives, in comparison with the standard theory, present a mixed picture. In some cases the alternative, depending on how it is construed, at least equals the standard explanation in this respect; while in other cases there is a clear inferiority, so far as that kind of theory has been developed so far. At this point I shall suppress these differences and assume, for the sake of argument, that all the alternatives, as so far developed, are markedly inferior to the standard explanation in explanatory detail. What implications should we take this to have for the truth of the standard explanation?

That depends on the first assumption listed earlier, "none of the other alternatives could be developed into something equally rich and detailed". Before determining what should be said about this assumption, I had better defend my supposition that it is a necessary condition of reasonably taking the standard explanation to be more likely to be true than its competitors.[10] The point is this. We are dealing with a difference between rival explanations that consists, not in the ways they are related to relevant evidence or anything else external that counts for or against their truth, but in the degree of their development. Now if a given explanation has been highly elaborated and is in good shape otherwise, while none of its possible alternatives have been, that is obviously a reason for working with the former, accepting it for practical and theoretical purposes, continuing to develop it, and so on. But is it a good reason for supposing the developed alternative to be true and its competitors false? I don't see that it is, unless we have sufficient reason to believe that none of the competitors could receive an equally elaborate development and still fare as well vis-à-vis the evidence. Unless we have such reason, we are allowing our judgment of truth to hang on what, so far as we know, is the historical accident that a great deal of effort has gone into pursuing this line of inquiry and theorizing, and little or none has gone into the other lines. And that is hardly a sufficient reason for supposing that the developed explanation corresponds to the way the world is.

Well, do we have sufficient reason for supposing that none of the alternatives (1) through (5) could be developed into something that exhibits an explanatory detail equivalent to that of the standard explanation? I cannot see that we do. The most basic point to make here is that it is, in general, impossible to predict, or set limits on, theoretical developments. Remember Auguste Comte's notorious example, in the first half of the nineteenth century, of an empirically undecidable question: the chemical composition of the stars! The general point is that to be in a position to predict successes and failures in theoretical developments, or in any field of creativity for that matter, we would have to have already achieved those successes (or failed to do so in such a way as to give sufficient reason for the impossibility); and so what is "predicted" would have come to pass in the act of prediction. It follows from this that the fact that a given theoretical development has not been attained, even after considerable effort, is

---

[10]I am grateful to Norman Kretzmann for inducing me to face this issue.

at best a very weak argument for the impossibility of such a development. But here we are not even in that position, since little or no effort
has gone into the development of the alternatives. I conclude that we
have no reason worthy of the name for supposing that the alternative
explanations could not be worked out with a richness of detail equal
to that of the standard explanation. Why suppose that "whatever
psychology and programming apparatus one could dream up for the
superbeings could not be as rich or testable as the theories of physical
science and neurophysiology"? As pointed out earlier, we can easily
understand why no such development has taken place, and this explanation has no tendency to show that the development is impossible. Even if greater richness in the sense that implies greater explanatory power, *where equal effort has gone into development,* is arguably a
token of greater probability of truth, that is not the case where the
developmental effort is grossly unequal. If you've just won the Heis-
man trophy, that doesn't show that you have more athletic ability than
I if I haven't done much to use what ability I have, and there is no
reason to think I couldn't develop into that good a football player if I
tried as hard as I can.

Next I turn to an interesting attempt by Laurence Bonjour (1985)
to compare standard and demon hypotheses. To be sure, he is not
comparing explanations of sensory experience. His explanandum is
one that grows out of his coherentist epistemology. He formulates it
as follows: "a system of beliefs remains coherent (and stable) over the
long run while continuing to satisfy the Observation Requirement"
(p. 171).[11] This can be taken as a coherence analogue of our explanandum plus the existence of a stable (and growing) system of beliefs
based on the experiences in question. But despite the important
differences in my and Bonjour's explanatory targets, his discussion of
the comparison of standard and "skeptical" hypotheses is well worthy
of consideration in the present connection; for the points he makes
would seem to fit our problem just as well, or just as ill. He makes a
number of points, many of them acute and well taken, but the bottom

---

[11]The Observation Requirement for a system of beliefs is that the "system must contain
laws attributing a high degree of reliability to a reasonable variety of cognitively spontaneous
beliefs" (p. 141). The term 'cognitively spontaneous belief' is a device for requiring the
justification of *perceptual* beliefs without actually coming out and saying so.

line is that the standard hypothesis is superior, in a way relevant to truth, in that its a priori probability is significantly greater. Here are the crucial passages.

> In considering the evil demon hypothesis, however, it is important to distinguish between two crucially different forms which it may take. . . . The first form postulates merely that there is an all-powerful evil demon who causes my experience. . . . without saying anything more about the demon's motives and purposes or about what sorts of beliefs he is inclined to produce; whereas the second form postulates in addition, as a part of the explanatory hypothesis itself, that the demon has certain specific desires, purposes, and so on, in virtue of which he will single-mindedly continue to produce in me, even in the long run, coherence-conducive observations. Hypotheses of the first kind are *simple demon hypotheses,* while hypotheses of the second kind are *elaborated demon hypotheses.* . . .
>
> A simple demon hypothesis, though it does provide an explanation of sorts for the long-run existence of a coherent (and stable) system of beliefs, fails to provide a very good explanation. . . . [S]uch an unspecified demon is capable of producing, and equally likely to produce, virtually any configuration of beliefs, and the simple demon hypothesis provides no reason at all for expecting him to confine himself to those which will fit coherently into my cognitive system. . . . This is scarcely a startling result, for it is obvious that it is elaborated demon hypotheses that provide the major skeptical challenge. . . . And against an appropriate elaborated demon hypothesis, the foregoing argument is entirely ineffective . . . an elaborated demon hypothesis of the right sort will make it extremely likely or even certain that my cognitively spontaneous beliefs will be coherence-conducive and hence extremely likely that my system of beliefs will remain coherent (and stable) in the long run. (Pp. 183–84)

However the elaborated hypothesis suffers from another crippling disability; its a priori probability is extremely low. This is because of the fact that wrecked the simple demon hypothesis, namely, the improbability of the explanandum given the existence of an evil demon. This improbability, says Bonjour is "internalized" by the elaborated hypothesis, with the result that it is inherently improbable.

> The unlikelihood that a demon would have just such desires and purposes (and that these would not change) seems no less great than the unlikelihood that an unspecified demon would produce just such observations. . . . [A] demon is capable of having *any* set of desires and

purposes, thus making the quite special set of desires and purposes which would lead him to produce a coherence-conducive set of observations equally unlikely. For this reason, the elaborated demon hypothesis is . . . extremely unlikely to be true. (P. 185)

One might well wonder just how Bonjour knows that a demon is capable of having any set of desires and purposes and (something which by no means follows from that,) that the likelihood of a demon's having just this set is extremely low. Perhaps he is in possession of an elaborated science of demonology to which the rest of us are not privy. But leaving that aside, I am afraid that Bonjour's procedure here, if generalized, would shoot down any attempted explanation whatsoever. Suppose I explain your giving me an unexpected present by the supposition that you are trying to influence me to give a favorable decision in a case in which you are interested. To make the case parallel, let's say that I have no independent reason for supposing that you have this motive. On Bonjour's principles, given that you are capable of having indefinitely many sets of desires and purposes, the antecedent likelihood of your having this set is infinitesimally small, and the hypothesis collapses before leaving the starting gate. Nor is the power of this principle confined to the invocation of personal agents as explanatory factors. Apart from any antecedent reason to the contrary, bodies are capable of attracting each other according to any of indefinitely many ratios of their mass and distance (not to mention other physical variables). Hence the antecedent probability of their doing so according to Newton's formula is vanishingly small, and we can reject Newton's law of gravitation a priori.

Indeed, it would seem that the standard explanation would also fall victim to Bonjour's argument, a point that he acknowledges.

The principal point at which the correspondence hypothesis [Bonjour's term for the standard explanation] seems to be vulnerable to an argument which would parallel the one already offered . . . is in its assertion that the cognitively spontaneous beliefs which are claimed within the system to be reliable are systematically caused by the kinds of external situations which they assert to obtain. It does not seem especially more or less likely a priori that there should be a world of the sort in question and that it should cause beliefs in some way or other than that there should be a demon which causes beliefs, leaving the two sorts of hypotheses roughly on a par in this respect. But if this is so, then it can be argued that the correspondence hypothesis is just as unlikely to be true on a purely a priori basis as are demon hypotheses. It is unlikely, relative to all the possible ways in which beliefs could be caused by the world, that they would be caused in the specific way required by the correspondence hypothesis. (P. 185)

But then why suppose that the standard explanation is superior to the demon explanation? For, according to Bonjour, the decision between them comes down to a priori probability, and, he has just admitted, they are equally bad off in this regard. And yet Bonjour proceeds to offer some rather weak considerations that are designed to show a slight edge of the correspondence hypothesis in this regard. I will cite only the second.

> Second, and more important, there is available a complicated albeit schematic account in terms of biological evolution and to some extent also cultural and conceptual evolution which explains how cognitive beings whose spontaneous beliefs are connected with the world in the right way could come to exist—an explanation which, speaking very intuitively, arises from within the general picture provided . . . rather than being arbitrarily imposed from the outside. (P. 187)

The idea will have to be that this evolutionary explanation is part of the hypothesis. If the claim were rather that the evolutionary explanation is correct, we would be involved in epistemic circularity again; for our reasons for accepting evolution stem from the very sources of belief we are trying to justify. But then it would be child's play to build something into the demon hypothesis as well that would explain why the demon prefers to produce sense experiences in the way we have them, rather than in some other way and rather than not producing any at all. Any appearance of a superiority of the "correspondence hypothesis" on this point stems from surreptitiously supposing ourselves to have independent grounds for accepting the evolutionary account, while lacking any independent grounds for suppositions about a demon. But, of course, those independent grounds are drawn from SP and other parts of our customary doxastic practices; to appeal to them would land us in epistemic circularity again. Thus, considerations of antecedent probability no more establish a superiority of the standard explanation than do considerations of simplicity, economy, depth, or detail.

The next attempt to establish the superiority of the standard explanation that I shall consider is that of Michael Slote (1970). Again, despite differences of detail, we may, for our purposes, take Slote to be engaged in choosing between alternative explanations of our explanandum. The alternatives he considers, in addition to the standard one, are our (1), the Cartesian demon, and the hypothesis that our experiences are caused by nothing, but "just happen". In seeking to show the superiority of the standard explanation he appeals to a principle of scientific methodology he calls the Principle of Unlimited Inquiry (PUI).

> a. It is scientifically unreasonable for someone to *accept* what (he sees or has reason to believe) is for him at that time an inquiry-limiting explanation of a certain phenomenon, other things being equal.
> b. There is a reason for such a person to *reject* such an explanation in favor of an acceptable non-inquiry-limiting explanation of the phenomenon in question, if he can find one. (P. 67)

"Inquiry-limiting explanation" is explained as follows:

> An hypothesis is inquiry-limiting for s as an explanation of certain phenomena at time t, just in case if s at t accepts that hypothesis and holds it to be the best and completest explanation of those phenomena available at t (and believes in the existence of those phenomena), he ensures the impossibility of his coming to have rationally justified or warranted belief (consistent with his other beliefs) in more and more true explanations of various aspects of or facts about the phenomena in question (for as long as he continues to accept that hypothesis as true and to believe it to be the best and completest explanation of the phenomena in question that was available at t). (P. 66)

In other words, an explanation is inquiry-limiting if its acceptance would foreclose the possibility of finding more explanation of facts concerning the subject matter in question. Slote argues for the acceptability of PUI as a principle of scientific inquiry on the grounds that science is essentially an attempt to explain as much as possible; and therefore any hypothesis that would imply that no further explanation is forthcoming will, by that fact, receive a black mark. PUI does not imply that no inquiry-limiting hypothesis could be accepted; the evidence for it might be so much stronger than its rivals as to force its acceptance on us, despite this drawback. The principle only holds that the acceptance of such an hypothesis is unreasonable, *other things being equal*, that is, if there is some alternative for which the evidence is at least equally strong that is not inquiry-limiting.

Slote then argues that, of our three competing explanations, only the standard explanation escapes the stigma of limiting inquiry. It is clear that it leaves us a clear field for seeking explanations of particular physical phenomena and of the ways in which particular kinds of experience are physically produced. And it is equally clear that the null explanation is maximally inquiry-limiting. If all that can be said of our experience is that it "just happens", we are indeed bereft of the possibility of explanation. The case with the Cartesian demon, however, is not so clear. Slote argues, rather persuasively, that the only way to seek explanations of why and how the demon produces our experiences as he does would be to ask him or at least to receive his messages on the subject. But since, by hypothesis, the demon is engaged in such a massive and long-continued deception, we would have no reason to trust what he says. Therefore (1) blocks any possibility of being rationally justified in accepting any further explanations of the way things go. Hence only the standard explanation is a rationally acceptable explanation of our experience.

If I had the space to do so, I would explore the possibility of finding out about the demon's motives and modi operandi by devices other than taking his word for it. However, I shall concentrate on what I take to be a more fundamental weakness in the position, namely, Slote's supposition that we can have adequate reason for accepting PUI independent of what we have learned empirically about the world and ourselves. I am quite prepared to acknowledge that PUI is a sound principle of *scientific* inquiry. But it by no means follows that the principle has any binding force on me if I am trying to decide whether to accept the existence of an external physical world or the reliability of SP. I can recognize the force of the principle for scientific inquiry because of my understanding of what science is, its goals and procedures. But I grasp all that only because I am au courant with an "external world" of physical things and processes, distributed in space, enduring in time, and investigated by scientists that form a sort of community. So long as all that is bracketed, as we must in this enterprise to avoid epistemic circularity, I can't draw on any of this empirical knowledge of the practice, goals, and norms of science. So long as I am behind this veil of ignorance I have no basis for accepting PUI or any other principle of scientific inquiry.

Slote recognizes this objection, and answers it as follows.

> Up to now I have been talking about actual scientific practice and about what "we" experience and have reason to think, in order to point up the validity of the principles I have used to show the (epistemic) reasonableness of believing in an external world. But one does not need to assume that there is an external world or that there are other persons in order to see the validity of these principles. . . .
>
> One can see their validity, I think, merely on the basis of the fact that one has had certain sense experiences of what seemed to be the activities of scientists in a real external world. Upon having such experiences one can see that certain principles of scientific inquiry make good sense and others do not, and if one then comes to wonder whether the external world and the on-going enterprise of science really exist, or even to believe that they do not, that will presumably not cause one to doubt that, *if* there is (were) a physical world with scientific activities going on within it, it is (would be) rational for scientists to adhere to those principles. (Pp. 87–88)

I will grant this point for the sake of argument, even though it is far from clear to me that I would be in a position to make rational judgments about what principles are normative for scientific inquiry if I were in doubt as to whether there is any such thing. But even granting Slote's claim, it does not follow that PUI holds for what I am doing in trying to decide between the various hypotheses in question as explanations of my sense experience. Deciding between those alternatives is *not* scientific inquiry. I am not in a position to engage in scientific inquiry until I have recognized a world of physical things and processes, and a community of persons with whom to join in investigating it. Since I am not doing science, why should the fact that PUI is normative for scientific inquiry have anything to do with what makes one or another choice more reasonable in this situation? More specifically, PUI is normative for science just because science is essentially an attempt to explain as much as possible; to go on from one explanation to another, cumulatively building a larger and larger theoretical system. But that's not at all what we are up to in trying to decide between different ultimate explanations of our sense experience, in order to determine whether sense perception is reliable.

There is no aim at, or expectation of, a cumulative sequence of explanations. We are not seeking to build up a comprehensive theory that will explain as much of the world as possible. That's just not the game here. We are engaged in the limited task of deciding between alternative fundamental metaphysical accounts of our sense experience. Inquiry-limitingness is in no way inimical to carrying out those aims.

Slote, like the other people we have been discussing, is quite limited in the range of alternatives he considers. Though he mentions self-generation he never really gives it a hearing; and there is no hint of panpsychism or of radically different physical explanations. If he were to consider these, he would find it much more difficult to pin the 'inquiry-limiting' label on them. I don't see why it should be deemed *impossible* to find out more and more about how I myself, or a world of Whiteheadian actual occasions, bring about our sense experience in the ways they do. Of course, there is no guarantee that we would actually keep finding out more and more; but neither is there any such guarantee with the standard account. PUI doesn't require of an acceptable explanation that it provide such a guarantee; it only requires that it does not *preclude* our continually adding to our explanatory accomplishments. And I can't see that self-generation or panpsychism can reasonably be thought to do so.

Finally, there is the point that Slote's appeal to PUI will show, at most, that it is more *reasonable* or *justified* to accept the standard explanation than to accept its rivals, in a sense of 'justified' that has no implications with respect to truth. It has no tendency to show that the standard explanation is true, or is more likely to be true than its rivals. That is, it has no such tendency unless we can show that non-inquiry-limitingness is a mark of truth; and I am at a loss to see how this might be done. After all, PUI was recommended, not as a way of determining what is true, but as a way of determining what is most useful in the realization of the goals of science (truth not being one of the goals under consideration there). Thus even if Slote's argument were successful, it would, so far as I can see, have no tendency to show that the picture of the world built up by relying on SP is correct, and hence no tendency to support the reliability of SP.

Finally, I shall look at a very interesting recent attempt by R. B. Brandt (1985) to give a noncircular rational justification of our basic doxastic practices, including SP. The program is announced as follows.

> I suggest we can identify a *policy* for belief-formation or belief-adjudication, composed of *imperatives* for the formation or appraisal of beliefs. . . . In saying we can "identify" a policy for belief-adjudication, I mean that we can pick out one policy (or a small family of policies), among possible policies, and show that it has a unique promise of recommending firm belief only in propositions that are true, firm disbelief only in propositions that are false, and intermediate degrees of belief only in propositions of types the frequency of truth among which is correspondingly great or small. (P. 12)

This is a lofty aim. And Brandt makes it clear that he aspires to carry this out without falling into epistemic circularity.

> Evidently one must produce some line of reasoning which does not beg any of the questions at issue—which doesn't just assume that the gaps may be bridged, which doesn't just assume that some person is right in his conclusions or his method. (P. 12)

Here is Brandt's proposed implementation of the program.

> I assume that we do have a conception of a world, or set of facts, independent of our beliefs about it, and that what it is for a belief to be true is that it represents correctly this set of facts. So what we want is a policy that will bring our beliefs as nearly as may be into correspondence with the set of facts.
>
> It could be that no policy can do this job. Indeed, it seems that the job can be done only if at least three conditions are met. The first is that there is enough lawful structure (it could be just statistical law) in the world so that a sampling of some piece of the world is a clue to the nature of other pieces of the world. The laws might change, and they might vary from one part of the world to another, but they must have enough range, and be simple enough, so that sampling and inference to other pieces of the world are reliable. The second condition is that there be some input from the world for a person, in the sense that there be some fact, distinct logically from a judgment about it, with which judgment about it can be directly compared, although not necessarily one about which judgment is infallible. Judgments about this fact may then be used as a starting point for inference, or a basis for appraisal of other judgments, say about laws or theories. Candidates for such a status, of course, are how we are being appeared to or what states of feeling we are in or what thoughts are running through our minds. If

there are such facts, then at any given time when we are awake we have some samples of the real world at our disposal, for use in theory-evaluation. It is not enough, however, that at any moment t we have some contact with the world, for purposes of theory-testing. For instance, even if there is a law of motion we cannot know what it is from just one observation; we need at least two samples. So, if we are to have a picture of the world at time t we need both input from the world at t and rather reliable information about inputs at other moments or about judgments about them made at this time. So there must be some sort of recording device, more or less reliable, for information about earlier inputs. So much at least must be the case if we are to attain reliable beliefs about parts of the world currently not observed. . . .

If these (and possibly further) conditions must be met if we are to form reliable opinions about the world, then we can see roughly what the policy with "unique promise" will be, and the sense in which it has "unique promise". Obviously the policy is going to direct that we form beliefs as if these conditions were met. It will direct us to form a consistent system of beliefs, necessarily including universal or statistical law-beliefs, since these alone will lead us beyond the content of our input. Second, it will direct us to incorporate in this system, at t, beliefs about the basic input at t; and usually if not always to strike beliefs incompatible with these (usually there will be a choice which beliefs to strike, whether law-like beliefs or auxiliary beliefs about conditions of observation and so on). Next it will direct us to incorporate the content of ostensible recollections about particular past experiences, or at least most of them. . . . [T]here is no simple rule informing us when a given one or more ostensible recollections should be rejected or accepted when faced with a complex of law-like beliefs supported by other ostensible recollections; I have no more precise suggestions than to follow the idealists and say one accepts the alternative least devastating to one's "intellectual world" as a whole. (Pp. 12–14)

I have said that this policy of belief-formation or appraisal has "unique promise" of leading to true beliefs (etc.). What this means is that the strategy is such that, *if* the world is such, or our relation to it is such, that true beliefs about it can be identified by a *systematic strategy*, then this strategy will lead us to them. Of course there *is* a world: "I doubt therefore I am" shows this. And given there is a world, if there are inputs which are samples, and there is a reasonably reliable record of what those samples have been, and if the parts of the world are related by laws that are not too complex (etc.), then if we form and test hypotheses involving laws on the basis of these samples, we shall be led to true beliefs, and away from false beliefs, about the world. These conditions may not obtain. If they do not, scientific reflection is idle. But if we want truth (and grasp its practical importance), we had better

hope for the best and adopt that policy which will generally lead to the
truth if truth is obtainable. (Pp. 14–15)

In other words, if and only if SP, introspection, memory, and inductive reasoning are reliable belief-forming practices do we have a hope
of forming mostly true beliefs. Hence, in view of our paramount aim
at attaining the truth and avoiding falsity, we have no choice but to
assume that these practices are reliable.

Brandt aims to avoid epistemic circularity. Instead of relying on
premises about what the world is like (which he would have had to
acquire by the use of doxastic practices he is seeking to validate), he
seeks to determine what the world would have to be like if we are to
have any chance of knowing it, and then considering what practices
would give us knowledge of the world if it is like that. Nevertheless, in
the end he does rely on the very practices he seeks to justify, albeit in a
subtle fashion. There are two points in his argument at which this
dependence can be spotted.

First, why should we suppose that Brandt's three conditions really
are necessary conditions of there being a policy of belief formation
that can be relied on to lead to the adoption of true beliefs and the
rejection of false beliefs? Let's look at the first condition: "that there is
enough lawful structure . . . in the world so that a sampling of some
pieces of the world is a clue to the nature of other pieces of the world".
To claim that this is a necessary condition for there being a reliable
policy of belief formation is to say that it is impossible for us to
consistently form true beliefs about parts of the world beyond our
experience unless parts of what we do experience are connected to
the former by lawful regularities. If that is the case, then we can
inductively discover those regularities and use them to extrapolate
from what we do experience to what we do not.

But surely this is not the only *logical* possibility. Surely there are
logically possible worlds in which a cognitive subject has *innate* knowledge of parts of the world it does not experience. And there are
logically possible worlds in which random guessing on the part of a
cognitive subject is a reliable method of forming beliefs about what
goes beyond its experience. So Brandt cannot reasonably claim that
his first condition is a logically necessary condition of our having a
reliable method of belief formation. Is it, then, necessary in some

other way? So far as I can see, the most that can reasonably be claimed on this score is that, *as human beings and the world in which they live are actually constituted,* it is not possible for us to reliably form beliefs about absent parts of the world unless there is enough lawful structure. But how do we know, or what justifies us in believing, that this is the way we and the world are? Obviously, by engaging in those very practices Brandt is seeking to show it to be reasonable to engage in (and hence, in practice, to take to be reliable). Epistemic circularity has crept in once more.

Second, suppose we grant Brandt his conditions and look at what he claims our policy must be if we are to maximize our chances of attaining the true and avoiding the false, if the world is as specified in those conditions. The second condition was that "there be some input from the world for a person, in the sense that there be some fact, distinct logically from a judgment about it, with which judgment about it can be directly compared. . . . Judgments about this fact may then be used as a starting point for inference, or a basis for appraisal of other judgments". Correspondingly, the second part of the policy "will direct us to incorporate in this system, at t, beliefs about the basic input at t". But the question remains: How do we identify "inputs from the world"? How do we distinguish them from other facts?

Brandt does not ignore this question. On the contrary, he informs us that "candidates for such a status, of course, are how we are being appeared to or what states of feeling we are in or what thoughts are running through our minds". In other words, he suggests that it is facts about sensory experience, feelings, thoughts, and other intro-spectable items that comprise what we should identify as "input from the world". But this is, of course, philosophically controversial. For one thing, Brandt is taking even sense perceptual "inputs" to be subjective or phenomenal in character; they amount to, as he says here, "ways of being appeared to", sensations of various sorts. But a direct realist would take perceptual input to consist of perceived facts about the external environment. Again, why limit the list in this way? Some, including the present writer, would regard one's awareness of God acting in one's life as an "input from the world".

The basic point here is that the mere requirement of *some* inputs from the world does not of itself tell us where to locate these. And if we ask Brandt why he thinks he has given the right list, I believe that a candid answer would be that what we know about ourselves, as cognitive subjects, in commerce with our environment, indicates that items of the sorts he mentions have the right properties to be regarded as inputs. They result from the impact of the world on us, and in such a way as to be reasonably reliable representations of that world. But how have we come to know that? By relying on just the doxastic practices Brandt is seeking to validate. And so epistemic circularity has once more marred the picture.

We have examined a number of attempts to show that the standard explanation of sense experience is superior to its rivals, and we have found them all wanting. We have by no means surveyed all such endeavors, but I fancy that the ones we have examined are among those most worthy of serious attention and that the failings they exhibit will be found to infect the others as well. At bottom the reason that so many acute philosophers have failed to do the job is the ban on epistemic circularity. When we are precluded from making use of anything we take ourselves to have learned from SP and whatever is based on that, we have little to go on in deciding which explanation of experience is mostly likely to be true. The various nonevidential criteria of explanations may be useful in certain choices between scientific theories, but in the present case they fail to pick a winner. It is beginning to look as if our assurance that the physical and social worlds are as we generally suppose them to be, and our confidence in SP, are too intimately connected, too *mutually* interdependent, to allow us to establish the former independently of the latter, and then use the former as our basis for the latter.

iv. Explanations of Our Success in Predicting Our Experience

We have been examining attempts to establish the reliability of SP by an appeal to the standard explanation of our sense experience or patterns thereof. The results have not been encouraging. No one has been able to mount an otherwise impressive argument without falling into epistemic circularity. There is a similar move, however, that is clearly superior to those we have been scrutinizing; amazingly enough it is not represented, or hardly represented, in the literature.

This new tack will involve a different explanandum. Instead of confining ourselves to sense experience and patterns therein, we enrich it
to include as well the following facts. (a) We use a certain physical and
social world scheme to conceptualize what we perceive, and we use a
certain procedure, SP, for forming perceptual beliefs. (b) By accepting the perceptual beliefs so formed and by developing systems of
belief on their basis, we are enabled to effectively predict *the course of
our experience*. We can often foresee with an amazing degree of accuracy what sensory experiences we will have at a given moment.
Furthermore, it is not just that we do this by relying on SP; that in
itself would be noteworthy, but hardly a decisive support for SP,
provided it were possible for us to do the same thing in other ways, for
example, by basing our predictions on observed regularities within
sense experience. The clinching point is that, so far as we can tell,
reliance on SP, along with memory and reasoning based on that, is
essential to the enterprise. We are simply unable to do the job just on
the basis of patterns within sense experience, the most obvious alternative. Even if there are enough stable regularities within sense experience to serve as a basis for the predictions, something we have
already seen ample reason to doubt, they would be too complex for us
to spot, store, and utilize. Think back on some of the points we were
making earlier. We can see from our familiar physical world scheme
that what prevents simple regularities in sense experience is that what
I experience at a given moment is a function not just of what I have
previously experienced but also of various details of the physical
setting, including the behavior of my body. Thus, if there are to be
any dependable phenomenal regularities, they will have to take into
account not only the immediately preceding phenomena but also
whatever sensory basis I have, in present and past experience, for
supposing the physical environment and my body at that time to be
so-and-so. The complexity this would assume is so staggering that one
cannot even form a plausible sketch of how it would go. Thus, it is not
only that these predictive achievements are made possible by the use
of SP; it looks as if we couldn't bring them off in any other way.

    The argument, then, is that far and away the best explanation for
this complex fact is that the scheme we use to bring off these predictions does fit the reality we perceive, and that the procedure we use to
form perceptual beliefs is a reliable source of belief. If our perceptual

beliefs are not mostly an accurate rendering of what we take ourselves to be perceiving, why is it that the forecasts of future experience we make on the basis of those beliefs should so often be borne out? It would be an incredible run of good luck. And so an inference to the best explanation assures us that SP is indeed reliable.

Let's note several crucial features of this argument.

(1) The reader will remember that an earlier argument from predictive success was dismissed as infected with epistemic circularity. This argument is specifically designed to avoid that disability. The earlier argument appealed to our success in making accurate predictions of external physical states of affairs. Hence it required reliance on SP to determine that those predictions are often correct. But this argument appeals to our success in predicting *the course of our sense experience*. It trades on our ability to anticipate that, for example, *it will be just as if we are seeing a maple tree in front of us*, rather than on our ability to anticipate that there will be a maple tree in front of us. Hence no reliance on the perception of external states of affairs is presupposed by the argument.[12]

(2) The argument avoids epistemic circularity at another point by merely appealing to the fact that *we use SP and associated doxastic practices involving the physical world scheme,* rather than itself working *within* those practices and thereby presupposing their reliability.

(3) It is clearly superior to the argument that is based on the standard explanation of patterns in our sense experience, superior in several respects. First, it doesn't require that any patterns or regularities can be found *within* sense experience. Its explanandum is rather the predictive success we attain when we abandon the search for such regularities and seek regularities in the objective states of affairs we take ourselves to perceive through sense experience. The

---

[12]At least, that's the way it appears on the surface. That appearance will be challenged by any view that, like Wittgenstein's, takes it that we cannot refer to sensory experience unless we are in possession of a public language that itself presupposes the reliability of sense perception. (For some consideration of this, see pp. 53–54.) For another example of views that don't allow this argument to escape epistemic circularity, see Jonathan Bennett's argument (1966, pp. 204–9) that it is impossible to have beliefs about one's own past sensory states unless one also takes them to be perceptions of objective states of affairs. More generally, we can say that this argument escapes epistemic circularity only if it is possible to think and speak about one's own experiences without already being *au courant* with a public physical world in a way that presupposes the reliability of sense perception. I shall assume that this is possible. I believe that to be the case, but even if I didn't I would assume it in order to give this line of argument a run for its money.

whole point of the argument is that by taking individual sense experiences as revelatory of external physical realities and connecting up our sense experiences through a detour into the physical world, we can do an incomparably better job of predicting those experiences than we could by looking for purely phenomenal patterns.

(4) The argument is also superior in that the desired conclusion, the reliability of SP, is built into the explanans that we are claiming to provide the best explanation. We don't, as we did earlier, have to argue from the explanatory claim to the epistemological conclusion of reliability by means of additional premises. Thus the argument, if sound, directly establishes the reliability of SP. This point is intimately connected with the previous one. It is because the explanandum concerns the predictive success we attain *when using SP* that the explanans features the reliability of SP.

(5) Finally, the argument is superior in that the favored explanation seems clearly to be preferable to its alternatives. What better explanation can there be of effective prediction on the basis of premises obtained by a certain doxastic practice, than the reliability of that practice? More generally, what better way can there be to explain the fact that the employment of certain methods and/or a certain view of things consistently leads to predictive success than to suppose that that method and that view of things puts us into accurate cognitive contact with the realities that are responsible for the events predicted? And, as just noted, the argument is greatly strengthened by the fact that there seems to be no other way in which we could achieve the same results. It is standard scientific procedure to reason in this way. If a particular account of the underlying structure of matter enables us to predict surface phenomena far better than any alternative account we have developed or can envisage, that is regarded as a decisive reason for taking that account to be correct. What better reason could we have, given that the underlying structure is not itself observable?

Recall that the proponents of earlier explanatory arguments were forced to try to show that the favored explanation was superior on grounds of simplicity or economy or detail, or on some other nonevidential ground. We saw reason to reject some of these claims of superiority, and in other cases we saw no reason to suppose that the superiority in that respect carried with it a greater probability of truth. But here, it would appear, we can dispense with appeals to such criteria. The argument is quite straightforward: we can effectively anticipate our experience by taking it as revelatory of the external world in accordance with SP; therefore, SP is probably quite reliable. What could be more direct or convincing?

v. Problems with the Argument

But despite these advantages, all is not clear sailing. The most obvious defect of the argument is the restricted character of the explanandum. If I am to avoid epistemic circularity, I cannot appeal to the success of other people in predicting their sense experience. Apart from reliance on SP I have no reason to think that there are other people, much less that they make use of SP and enjoy predictive success thereby. Hence the explanandum is restricted to *my* predictive success when using SP. How can we hang so enormous a conclusion on so slender a thread? Leaving aside what is based, at least in part, on SP, I might, so far as I can tell, be the only person in existence, or the only one who uses SP, or the only one who enjoys such success from employing it. Let's concentrate on the last of these possibilities. If that were the case, if all the other billions of human beings that have existed have failed to reap such results from the use of SP, we could hardly conclude that SP is reliable. It would not be an incredible run of luck for one out of billions to experience predictive success. Of course, I am not suggesting that this is actually the case. Rather, I am arguing that if I am to avoid epistemic circularity I can't assume either that it is or that it isn't. Behind my veil of ignorance I cannot suppose myself to know anything at all about the matter. Hence the argument is an exceedingly weak one, since it cannot rule out a possibility that, if realized, would render it impotent. So long as we have nothing but one case to go on we are in no position to choose between *SP is reliable* and *I am an exceptionally lucky case.*

It may be thought that this objection simply amounts to pointing out that no inductive argument is conclusive, that any such argument can be overturned by unexamined cases. Not so. This is not an enumerative induction at all. It is an argument to the best explanation. And the point is that the narrow compass of the explanandum prevents us from showing that the suggested explanation is in fact the best one.

Interestingly enough, if, contrary to the above criticism, I could succeed in showing that SP is reliable in my case, there is a route from there to the conclusion that SP is generally reliable. For if SP is reliable in my case, then the beliefs I get from it are mostly true. And unless I have some special reason for supposing that beliefs about other persons—their existence, their employment of SP, and their successes in doing so—are mostly to be found in a minority of false beliefs yielded (in part) by SP, then it follows that what I believe about other persons is mostly true. That is, it follows that there are lots of other people who have had the same kind of success with SP. And so the argument for the general reliability of SP would go through after all.[13] But that is all on the supposition that the argument is sufficient to establish the reliability of SP in my own case. And that is precisely what I have just been contesting.

I feel that the above is sufficient for rejecting the claim that I can become justified in accepting the reliability of SP by noting that it best explains my success in predicting my sense experiences by using SP. Nevertheless, I will set the above objections aside for the nonce and consider how matters stand with respect to another component of the argument, the claim that the predictive success cannot be attained in any other way. We have seen that, so far as we can tell, we are incapable of taking a purely phenomenal route to effective prediction. But we have not yet explored alternative objective routes, via the demon or panpsychism or whatever. Why shouldn't we employ one of those alternatives in the following way? The demon, let's say, directly produces our sense experiences, together with inclinations to employ SP and other components of the physical world scheme, in such a way as to lead us to make accurate predictions of our sense experience. In other words, the demon sets everything up in just the way it would be set up if there really were a physical world with which we are in effective cognitive contact via SP. And we can envisage an exact parallel for the panpsychist and self-generation hypotheses. Why isn't

---

[13]I am indebted to Norman Kretzmann and students in the seminar he gave on my book *Perceiving God* (1991a) for suggesting this line of argument.

this just as good an explanation of our predictive success as the standard hypothesis?

There is a good answer to that question. These alternatives are riding piggyback on the standard theory. They haven't been developed independently; that would require, for example, developing a theory of demonic psychology on the basis of which we could predict what sense experiences we would have, given the current and past states of the demon's psyche. We don't have anything like that, nor do we have the analogues for the other alternatives: for example, a theory of the internal structure, operations, and interactions of Whiteheadian actual occasions that would enable us to predict future experiences on the basis of detailed descriptions of the external environment in those terms. Instead we just latch onto the way all this has been developed in detail for the standard theory, and lamely say that, for example, the demon produces experiences so that things will come out this way when we employ SP. This means, first of all, that these alternatives are less simple, for they contain, if fully spelled out, all the complexity of the standard account, plus something extra.[14] And, worse, the alternatives suffer from being ad hoc. There is no serious theoretical motivation or justification for tacking the demon or self-generation or panpsychist hypothesis onto the detailed standard theory.

But not so fast. Let's grant that these alternative explanations, as they have been developed and *as far as we are presently able to envisage further developments*, are markedly inferior to the standard explanation. But why suppose it to be impossible for us to develop one or more of them on their own (not piggyback on the standard theory) in such a way as to be at least equal in explanatory power with the standard view? In discussing the argument from the explanation of our sense experience in section iii we noted that it is very difficult to

[14]Recall that a similar charge against the demon theory was rebutted earlier, on the grounds that the demon need not be using the standard theory as a model. But that was before we built *our predictive success when using the standard theory* into the explanandum. Since we have done so, reference to the standard theory in the alternative explanations is unavoidable. Merely pointing out that the demon directly produces our sense experiences does nothing to explain our success in predicting our experiences when we use SP. In the absence of a usable demonic psychology, if we are to explain that via the demon, we have to enrich the explanation to include the intention of the demon to produce those experiences in just the way they would be produced by physical causes according to the standard explanation.

show that a certain theoretical development, or any other creative advance, is impossible for us. In the nature of the case, we cannot hope to anticipate future theoretical advances. Hence, the fact that powerful theories have not been developed along certain lines up to this point is but a weak reason for supposing that it is impossible that they should. And, as noted earlier, this is especially true when, as in the present case, little or no effort has gone into the attempt. Thus, since the argument depends on the premise that we *cannot* achieve equal success in predicting the course of our sense experience on some basis other than SP, it is not in a very strong position.

But suppose that it is impossible for us to develop independently an equally powerful alternative theory to use in the prediction of the course of our experience? Would that really be a strong reason for supposing that SP is generally giving us the truth about reality? Not necessarily. After all, the fact, if it is a fact, that we human beings cannot mount an effective prediction of our experience with, for example, purely phenomenal means is at least as much a fact about us as about what the rest of the world is like. And the same is to be said for our supposed inability to do the job independently in terms of a demon or a panpsychist scheme. To be sure, the fact that the standard theory has been developed in detail and these putative rivals have not, is itself a fact that requires explanation; but our cognitive successes and failures in dealing with reality are due to an interaction between us and the subject matter. If either factor were sufficiently different, the outcome would be different as well. If the course of our experience were such as to present sufficiently simple lawlike phenomenal regularities, we might never have so much as thought of the standard view, and even if we had, it would have had much less to recommend it. Even given the internal irregularity of our sense experience, the nature of the physical world might not have been such as to permit the development of the relatively simple physical object scheme we actually have. From our side, if our cognitive powers had been quite different, the epistemic situation would have been correspondingly different. If we had intuitive powers of the sort ascribed to angels by Aquinas, we could have known about the physical world in a way unmediated by sense experience; and the issue with which we are dealing would not have arisen. Again, we might have been without our innate tendencies to form perceptual beliefs about an external physical world, in which case what is in fact the standard external world theory would perhaps never have occurred to us. I am engaging in this science-cum-theological fiction to sharpen the point that when we try to assess the ontological and epistemological significance of the fact that we do an impressive predictive job with the standard scheme, and can't do so with any alternative, we are faced with the question of how much this is due to the subject matter and how much it is due to our cognitive powers. It is only if the weight of emphasis falls on the former that we are entitled to the explanation in terms of the way the world is.

In addressing this issue it is highly relevant to consider what is possible along these lines for a cognitive subject of a higher grade than ourselves. Suppose it were true that a cognitive subject with much greater computing capacity—one who could note, store, combine, reason, and otherwise work with much more complicated formulations—could do at least as good a job of predicting sense experience as we do with SP while working purely with phenomenal patterns? Wouldn't that cast doubt on the claim that our inability to do the job with the purely phenomenal regularities and our ability to do it with the standard scheme is a good reason for thinking that the standard scheme captures the determinants of our experience? Admittedly this state of affairs would still be compatible with the truth of the standard scheme; but the current argument for that scheme would be greatly weakened, since our cognitive situation would be revealed as the result of limits on our powers, whether reality is as we ordinarily think it is or not. Contrariwise, if it is impossible for any subject, however complex, to do the job in phenomenal terms or in terms of any other alternative scheme, that would strengthen the reasons for thinking that our success betokens the accuracy of the scheme employed. The same point holds for the other alternatives. If it is possible for some cognitive subject to develop the demon hypothesis or panpsychism independently as a basis for successful prediction, then the fact that we have actually done so with the standard scheme but not with these others looks to be due to our limitations or accidents of our history, rather than to constraints that stem from the nature of things. We should not rush into ontological or epistemological conclusions from our cognitive successes and failures without further consideration of how they are best accounted for.

This is a good time to remind ourselves that this inquiry is being conducted under a realist assumption that things are as they are regardless of how we think of them or what we believe them to be, and regardless of what way of thinking of them is most approved by our epistemic principles. It is within this realist context that we are confronted with the question of whether beliefs, theories, and explanations that come out winners on accepted criteria do accurately represent the ways things are in themselves. On a nonrealist view, according to which "the way things are" just is "what suppositions about things pass the standard tests", there is no room for such a question. There is nothing that might be otherwise when our best criteria indicate that things are so-and-so. No doubt, realism makes life more difficult; but it does have the advantage of being true.

Thus, the crucial question is as to whether it is possible, either for ourselves or for other cognitive subjects, to successfully predict sense experience on bases other than the standard theory. In considering this question let's declare out of bounds appeals to an essentially omniscient God, Who can predict anything infallibly on no basis whatever. Since the predictive success of such a being is equally compatible with any view about the subject matter, it is irrelevant to our present concerns. Confining ourselves to finite subjects of limited capacity, it would seem that we have little enough to go on. How can we tell what is possible, for some finite cognitive subject, by way of theoretical-cum-predictive achievements? As just pointed out, it is notoriously impossible to predict actual future theoretical developments even by human beings. How much worse is our position to determine whether it is *possible* for radically superior cognitive subjects to do an equally effective job of predicting sense experience without working through the standard physical world scheme. We can be assured that none of us is at present able to do so, and that the motivation to develop such an ability is lacking. But to have a solid basis for attributing our predictive successes and failures to the fact that the scheme we work with is basically *correct*, we would need good reasons for thinking that no fundamentally different scheme would do as well for any finite cognitive subject. And we have no such reasons.

It may be felt that I am unfairly placing the burden of proof on my opponent. "Why", she might say, "do I have to show that it is impossible for any cognitive subject to predict sense experience on a different basis? Why isn't it up to you to show that this is possible, or at least to give a reason for taking the possibility seriously? Why do I have to assume the burden of eliminating every hare-brained scheme that anyone mentions?" Well, where the burden of proof lies depends on who it is that is making a claim. The claim under discussion here is that SP is reliable, and the argument for that claim is that our success in predicting sense experience on the basis of SP and associated practices is best explained by supposing SP and the associated practices to be reliable. That argument provides adequate support for the claim only if it is adjoined to an adequate argument that the success in question, and our failure to do the job in any other way, reflects the accuracy of the practices employed, and not just our cognitive limitations. So the reason it is incumbent on my opponent to show that a more powerful cognitive subject couldn't do as good a job on a quite different basis is that this is needed to shore up her argument for a claim she is putting forward.

Indeed, it is not just that we find ourselves unable to show that it couldn't be done otherwise; there are positive reasons for supposing it can, though these could hardly be deemed conclusive. When we think about the project of proceeding solely on the basis of phenomenal regularities, what appears to be the main bar to carrying this through? The regularities we can actually formulate are, as noted earlier, too local, too much at the mercy of shifts in the external environment; they have no claim to be regarded as lawlike. When we consider how this might be overcome by further elaboration, including in the antecedents enough sensory evidence for the physical environment being just right for the production of the consequent, things soon become too complex for us to handle, we get dizzy, and we give up. But this gives rise to the suggestion that *a subject with sufficient cognitive capacity would not have to give up at this point, and could proceed to put in enough phenomenal detail to ensure that the phenomenal consequent is as specified.* At least, *so far as we can see,* this is a live possibility for a subject with sufficient computational capacity.

Moreover, even if no one could do it differently, there are considerations that can legitimately inhibit us from inferring that the subject matter must be in accordance with the way we have to think of it. There are many cases in which a certain scheme serves us well, predictively and explanatorily, but where there are reasons to think that the scheme does not accurately capture the intrinsic nature of the subject matter and that another scheme does so, or could if developed properly, even though we are unable to work out the latter approach in detail.[15] First, think of the much maligned "folk psychology". We, not only we humble laypersons but social scientists as well, work for the most part with our familiar scheme of perceptions, beliefs, desires, intentions, attitudes, emotions, and other intentional psychological phenomena, in seeking to explain and predict behavior. And it works reasonably well. It doesn't give us the degree and precision of prediction that we have in chemistry or physics, but it gives us much more than we get with any other approach. On the other hand, various philosophers and psychologists have given reasons for thinking that this familiar folk psychological scheme is radically defective in ways that debar it from serious consideration as a literal account of the way things are with us. I won't try to go into those reasons here, but they include such things as the ontological baggage involved, the assumptions we must make to determine the intentional psychological states of others, the limitations on the development of theories involving such concepts, and the requirements of materialism. Still less will I undertake to assess these reasons. Suffice it to say that some case has been made for the thesis that what we human beings are actually like is captured only in more hard-headed physiological concepts. I put this forward as an example of a not obviously false and not obviously ill-supported claim that the predictive success we enjoy with a scheme is not a sufficient support for its literal accuracy.

Again, consider the frequent recourse to idealization in science. Physicists and engineers work with notions of perfectly rigid rods, frictionless planes, and the like, even though they realize there are no such things. By proceeding in this way we can spell out formulas that are simple enough to use for prediction and design, formulas to which many actual situations are close approximations. Likewise, in many applications it is more useful for predictive and applied purposes to use a simpler approximation to the truth, like Newtonian

---

[15]I may as well acknowledge in advance that some of my examples will be philosophically controversial, but I am not using them to establish an actuality, only to illustrate the way in which we can have grounds for suspecting the literal truth of a scheme that serves us much better in prediction and explanation than any alternative available to us.

physics, rather than a more complex true (or closer to being true) theory like relativity theory. In all these cases there is widespread agreement that we use the scheme we do because of our cognitive limitations, that the reality we are dealing with is not that way, and that a more powerful intellect could do an equally effective predictive job while representing things as they really are.

Thus, the fact that we can effectively predict our sense experience by using SP is, at best, a weak reason for supposing SP to be reliable. Hence, even the strongest epistemically noncircular argument for the reliability of sense perception falls considerably short of giving adequate support to that thesis.

At this point I can imagine a reader feeling that we have been applying exaggeratedly high standards to this explanatory argument. "After all", the protest might go, "you have admitted that the standard explanation for our success in using SP to predict our experience is markedly superior in explanatory force and detail to all rivals in the field. But that is a basis on which we regularly accept explanations and explanatory hypotheses and theories. If a certain explanation of my car engine stalling is clearly superior to any existing rival in this way, we have no hesitation in taking it to be correct. If a certain explanation of the presence of water on the floor of my basement is clearly superior in these respects to any alternative we can think of, again we do not cavil at taking it to be the correct explanation. Suppose, in this latter case, that the live alternatives are (1) leakage from the hot water heater, (2) seepage through the walls, (3) overflow of a reserve well, and (4) water from a heavy rain coming into windows. But we can eliminate (4) because there has been no recent heavy rain; and (2) and (3) fall by the wayside because it has been an unusually dry summer, and the ground is not at all wet. That leaves (1) in possession of the field, a position strengthened by the fact that the water is near the water heater. Now, of course, there are many other logically possible explanations. The water could have been created ex nihilo by God as a sign of something or other. Gremlins or fairies could have lugged it in under cover of darkness. Some fantastically improbable concatenation of quantum phenomena could have resulted in an assemblage of water at this point. But we don't take any of that seriously enough to give it a second (or even a first) thought. We don't bother our heads about whether one of these outré explanations could be developed into a serious rival to (1). We don't think it worth our time and effort to look into the matter. We have more than enough on our hands in dealing with explanations that are live possibilities, given what we (think we) know about the subject matter, without chasing after logical possibilities that we have no reason to take seriously. If all but one of the live possibilities can be ruled out, that settles the matter. The remaining live candidate embodies the true account of what is responsible for the phenomenon; and that's that. The fact that radically different explanations might conceivably be developed into serious contenders is too remote from our current view of things to be worth consideration."

The same point can be made concerning our standard practice of evaluating scientific explanations. There too our current view of the subject matter provides us with a basis for distinguishing between those conceivable explanations that are live possibilities and those that are not. There is no doubt but that these lines of demarcation change with the development of a science. At a certain point in the development of modern physics only mechanical explanations were countenanced. Later electromagnetic explanations were added, and still later explanations in terms of the curvature of space-time. Nevertheless, at any given time there will be a fairly sharp line between conceivable explanations that are live possibilities and those that are not, a line that is based on the currently dominant views of the subject matter of the science. The standard procedure in seeking an explanation of a current phenomenon is to consider the competition only between those alternatives that count as live possibilities by current standards.

All this suggests something similar as the reasonable way to proceed in evaluating alternative explanations of our predictive success with SP. If one explanation stands out as the only survivor from among the live alternatives that has been sufficiently developed, then that settles the matter. We don't have to show that no conceivable alternative could possibly be developed into an equally strong contender before we are fully justified in taking the former to be the correct explanation.

This reaction certainly deserves careful consideration. But on careful consideration I find that the analogy does not hold up. One plausible reason for that judgment is that, in the examples just given, we are only trying to decide what explanation to accept for working purposes—practical or theoretical as the case may be. In the basement case, we are simply picking a hypothesis to act on; we are trying to decide what measures to take to prevent future water incursions; we are trying to decide where to apply our efforts. In the scientific case, we are deciding what explanatory hypothesis to accept as a basis for further experimentation and theoretical development. In none of these cases is it a question of commitment to something or other as the *true* or the *correct* account of what is responsible for the phenomenon. But our problem here precisely has to do with truth. The explanatory claims were put forward as a way to establish the *truth* of the thesis that SP is reliable. We are not just interested in whether we are well advised to accept the reliability of SP as a working hypothesis or as a basis for action. If we can't determine whether SP *is* reliable, as a matter of hard fact, then, as I shall be suggesting in the final chapter, it becomes of central importance to determine whether assuming SP to be reliable is the most reasonable course to take from a practical standpoint. But in the main body of this essay we are considering whether it is possible to show, without epistemic circularity, that SP is reliable. And exhibiting the practical benefits of assuming this does not amount to establishing its truth.

However, I don't want to attack the analogy on these grounds. I do not agree that our choices between competing explanations in science and in everyday life are not concerned with truth. In deciding how to keep unwanted water out of the basement in the future, we need to know what was truly responsible for this pool of water on the basement floor. And in deciding what explanation of a given phenomenon to adopt for further scientific development, we want, so far as in us lies, to adopt a hypothesis that makes explicit what was in fact responsible for the phenomenon in question. Hence, I do think that our standard procedures for evaluating explanations aim at picking the true explanation. If the analogy holds up on other grounds, we are justified in proceeding in the ways just illustrated when seeking the best explanation of our predictive success when using SP.

But that's the rub. There are other, crucial respects in which the analogy does not hold up. An essential component of the standard way of evaluating explanations, as I have been presenting it, is the discrimination between real possibilities for explanation and mere logical possibilities. That discrimination, as I pointed out, is based on what we know about the subject matter in question. We often know enough about the kind of explanandum in question and its relations to other things to enable us to make a list of the ways in which it might be brought about, or of the kinds of factors that might be responsible for it. I know enough about the intrusion of water onto basement floors, and enough about this basement in particular, to have a pretty good idea of what can and what cannot bring about such an outcome. A physicist knows enough about the expansion of a given quantity of matter to have a pretty good idea of what could set off and maintain the expansion of the universe. What this means is that we are only able to evaluate explanations in the usual way, where we have the kind of knowledge that is needed to put us in a position to make well grounded judgments as to what can and what cannot be responsible for this kind of explanandum.

And that is precisely what is lacking in our case. Remember that we are looking for a way of establishing the reliability of SP that does not involve epistemic circularity. That means that we cannot appeal to anything we take ourselves to know, where that knowledge rests, even in part, on SP. Hence our situation is radically different from the one we are in when we look for explanations *within* our familiar basic doxastic practices, free to take for granted that these practices are reliable and hence that the beliefs we acquire by engaging in them are, by and large, true. In this latter case, we can draw on an immense amount of knowledge of water, basements, physical forces, and so on. This is what enables us to effectively distinguish between possibilities that are live and those that are not, and hence enables us to achieve an assured result by eliminating all except one of the live alternatives. But where the terms of the enterprise prevent us from utilizing what we have learned from SP, together with associated practices, we are bereft of virtually all reasons for distinguishing, among logically possible explanations, between those that are real possibilities and those that are not. It is natural for us to suppose that the standard explanation of our success with SP could very well specify what is responsible for it, while the others we have been considering could not. But in making that judgment we are relying on what we take ourselves to know about human perceptual, and other cognitive, activity in commerce with the physical and social environment. Given what we think we know about this, it is enormously more plausible, to say the least, that this predictive success is due to the reliability of SP than to the machinations of a Cartesian demon or the dispositions of Leibnizian monads. But this alleged knowledge is based, at least in part, on what we take ourselves to have learned from SP. Take away the information we get from SP about the immediate environment, and all this knowledge dissipates like the phantoms of a dream. Hence, in the radical situation in which the present inquiry is being conducted there is no basis for taking the standard explanation to be more of a real possibility than explanations in terms of Cartesian demons or self-generation. Once we eschew epistemic circularity, we have no basis for that preference. And that means that we are simply not in a position to follow our usual procedures for the evaluation of explanations.

It may seem that I have strayed from the challenge I set out to answer. Let me reconstruct that challenge. It concerned my last reason for denying that the explanation of the success of SP in predicting the course of sensory experience in terms of its reliability provides an adequate reason for supposing it to be reliable. That reason was that even though the standard explanation, in terms of the reliability of SP, is miles head of any competitors in terms of richness of explanatory detail, that will not give us a sufficient reason for taking it to be the true explanation unless we have sufficient reason for supposing that no alternative explanation can be developed to a point of equal explanatory richness; and we have no such reason. The challenge was that our standard procedure for ranking competing explanations shows that it is unreasonable to lay down such extreme requirements for justifiably accepting one explanation over against its competitors. For when proceeding as we standardly do, we regularly take a given explanation to be definitively established, even though we have not shown that no possible alternative explanation could be developed so as to do as good an explanatory job. We just do not take that to be necessary for determining what the true explanation of something is. My response to that challenge was that this standard procedure depended on our capacity to use a rich body of background knowledge to distinguish between those logically possible explanations that are live possibilities and those that are not. And in the present case, we are, because of the ban on epistemic circularity, debarred from making use of any such body of background knowledge, and hence from distinguishing live from nonlive possibilities among logically possible explanations. And that means that the criteria we employ in situations where such distinction is possible have no application to our problem.

Against that background we can return to the present stage of the controversy. "Even granting the last point", my challenger might say, "it still remains that the standard explanation is the only one that has in fact been developed in such richness of explanatory detail. And that remains a highly significant difference between that explanation and any of its possible alternatives. Hence, even if, going on what we are allowed here, we can draw no distinction between those possible explanations that are relatively live and those that are relatively dead, even if, so far as we can tell within our self-imposed restrictions, they are all equally live, it is still the case that one of those possible explanations is far superior to any of the others in the systematic detail to which its explanation has been developed. And that still remains a conclusive reason for supposing that to be the correct explanation. The fact that we can't marshall background knowledge to justify excluding its competitors from the class of live possibilities does not diminish the force of this point."

Ah, but it does! Just because we are not in a position to dismiss the alternatives as not being live possibilities for bringing out what is responsible for the explanandum, we cannot settle the matter just by noting that only one explanation has undergone an extensive systematic development. If we were in a position to brand the Cartesian demon, etc., as not being serious possibilities, we would be in that position. We would be in the same position as with the water on the floor of my basement, where I felt comfortable about ignoring gremlins and pixies as possible explanations of the presence of that water, even though I couldn't show, or hadn't shown, that those explanations could not be developed into something as impressive as the *leak in the hot water heater* explanation. But since here we are not and cannot be in such a position, we cannot dismiss these undeveloped possibilities without having sufficient reasons for supposing that they could not be developed into equally impressive explanations, either by us or by cognitive subjects with greater powers. And, as I have argued, we lack such reasons. Hence the earlier judgment stands. Though the standard explanation is unquestionably the only one that has been developed with a signal richness of detail, that does not provide us with an adequate reason for regarding it as the true explanation, and hence it does not provide us with an adequate reason for taking SP to be reliable.

vi. How Widespread Is the Circularity Problem?

I have examined a large number of attempts to show in a noncircular fashion that sense perception is a reliable guide to the external environment. I have, in fact, examined all the more impressive-looking attempts known to me. None of them have survived scrutiny. Unless and until someone comes up with a more successful alternative we will have to conclude that we are unable to give a noncircular demonstration, or even a strong supporting argument, for the reliability of SP. And that raises the urgent question of what attitude it is reasonable to take toward this way of forming beliefs, given the impossibility of an otherwise cogent noncircular argument for its reliability.

Before grappling with that question, let's consider how widespread this situation is with respect to our most fundamental belief-forming practices, such as memory and inductive inference. I take our position vis-à-vis SP to be representative of the epistemic status of all our basic sources of belief, at least so far as our ability to carry through a fully satisfactory demonstration of their reliability is concerned. To show this I would, in each case, have to critically examine the most promising attempts to give a noncircular demonstration of reliability, as I have just done for SP. In most cases, the length of the discussion would be considerably shorter, since, except for induction, the volume of literature on the topic is much smaller. But even so, carrying out this task would swell the present essay far beyond its intended bulk. Hence, I will confine myself to reminding the reader of some well known considerations that suggest that for one or another of these sources the prospect of a noncircular demonstration of reliability is not a bright one.

First, consider induction, construed narrowly as the inference of a generalization from a number of instances (induction by simple enumeration). We heat numerous samples of lead at sea level and ascertain in each case that the substance melts at (approximately) 327 degrees C. We conclude that lead always melts at sea level at (approximately) 327 degrees C. Since Hume's classic discussion in the *Treatise of Human Nature*, it has been generally recognized that any otherwise effective attempt to show that this procedure is reliable (yields mostly true beliefs from true premises) will be circular. Any track record argument will itself be an induction by simple enumeration, and so, in using it to establish reliability, we are, in practice, assuming that form of argument to be reliable. Where else can we turn? It is most implausible to suppose that the general reliability of this mode of inference is self-evident or otherwise discloses itself to rational intuition, nor have attempts to deduce it from non-inductively justified beliefs been successful. Attempts have been made to establish the principle as the best explanation of something or other, but they have not been particularly plausible. There have been many attempts to bypass the issue of reliability by arguing, for example, that it is part of our concept of rationality, or of good inference, or of showing or establishing something, or . . . that the positively evaluative label in question applies to inductive inference; but even if these attempts are successful, they most definitely do not show that induction by simple enumeration is reliable. Thus, our situation here would seem to be closely analogous to our situation with respect to SP.

As for deduction, it quickly becomes obvious that anything that would count as showing that deduction is reliable would have to involve deductive inference, and so would assume the reliability of deduction. Just try it. For example, for the case of the propositional calculus, we can demonstrate the reliability of any inference form by truth tables. But doing so is itself a case of deduction. We might try an induction from a number of cases in which true premises deductively yield a true conclusion, but for this to be noncircular it would have to be the case that our justification for these attributions of truth rested nowhere on the use of deductive inference, and the prospects for this are not rosy. More specifically, to the extent that we choose our sample with that restriction in mind, the inductive inference is correspondingly weakened by virtue of the truncated character of the sample.

Memory presents an interesting case. It looks at first sight as if we can develop a track record argument for the reliability of memory without using memory to get any of our premises. For we do have ways of determining whether p was the case at some past time other than remembering p to have happened. To determine whether our current TV set was delivered on August 6, 1991, we can look at the delivery slip, consult records in the store, and so on. But the delivery slip is good evidence only if this is the slip that accompanied that set; and I have to rely on my memory to assure myself of that, or else appeal to another record with respect to which the same problem arises. More generally, the reliability of records, traces of past events like the disorder in the wake of a party, and so on, is itself something for which we need evidence, and to marshall that evidence we will have to rely on memory at some point. I will have to remember that in the past, parties like that one have left the house in some disarray; or else I will have to consult diary entries that indicate this in various instances, in which case I have to remember that I made that entry as a record of what I had observed to happen; or I will have to draw on the experience of others, in which case I will be relying on their memories at some point. And so it goes.

Again, consider introspection. Here we would seem not to be in as desperate a situation for independent checks on the subject matter as we are with sense perception and the environment. Don't we have third-person ways of determining what a given person is thinking, sensing, or feeling at a given moment? To be sure, these third-person resources are much more limited than the first person's introspective access. They don't give us nearly as much information, and what they do give us is much less certain, liable at every point to be overturned by the subject's own sincere report of her thoughts and feelings. Nevertheless, so far as they go, don't they constitute an independent way of determining the truth value of introspectively derived beliefs? But are they completely independent? Isn't our confidence in these external indications of conscious states ultimately based on correlations with first-person reports? Apart from such correlations do we have any sufficient reason to credit such third-person manifestations? If the answer to these questions is in the affirmative, then third-person indications of conscious states have basically the same status as nonperceptual indications of perceivable physical facts; they depend for their epistemic status on connections with the more basic perceptual or introspective access to the subject matter. And so they don't provide a wholly independent check. This conclusion will be contested by those who suppose some external evidence for conscious states to be "criterial", to have their status guaranteed by the concepts of the conscious states in question. I find this Wittgensteinian position quite unpersuasive, but I can't properly go into that here.[16]

These sketchy remarks suggest that the situation we found to hold for SP obtains with respect to all our most fundamental belief-forming practices, whether the input is taken from memory, introspection, rational intuition, or reasoning of one kind or another. If so, we are confronted with whatever problems are thus engendered over the whole range of our cognitive life. But suppose we are mistaken in this. Suppose that sense perception or introspection or one or another sort of reasoning can be noncircularly shown to be reliable. Suppose that for one or another of these belief sources there is a successful argument for its reliability that takes its premises exclusively from other sources. Even so, we would not have escaped the necessity of dealing with the problem of how to regard doxastic practices we all engage in without being able to show that they are reliable. Let's say, contrary to our contentions here, that we can noncircularly establish the reliability of SP. Suppose that this argument appeals only to rational intuition and deductive reasoning; the argument deduces the reliability of SP from self-evident principles. What about those practices? Consider one of them—rational intuition. Can we mount a noncircular proof of its reliability? If we can't, we have the same problem at a second remove. If we can, then if that proof depends on using SP we are involved in a very small circle. If we do not have to use SP, let's consider the practice we do use. Can we give a noncircular proof of its reliability? If not, our original problem has been postponed to this point. And so on. We are faced with the familiar dilemma of continuing the regress or falling into circularity. Whatever the possibilities of a noncircular proof of reliability for one or another source, if we pursue the question far enough we will either (a) encounter one or more sources for which a noncircular proof cannot be given, or (b) we will be caught up in circularity, or (c) we will be involved in an infinite regress. Since the number of basic sources is quite small for human beings, we can ignore (c), and for the same

---

[16]For some discussion of a criterial approach to the epistemology of sense perception, see section iii of Chapter 3.

reason any circle involved will be a small one. Thus, in practice we can say that, whatever the details of our epistemic situation, either there are some doxastic practices for which we cannot give a noncircular demonstration of reliability, or in giving such demonstrations we involve ourselves in a very small circle. Thus, whatever the details, we cannot escape the question of what attitude it is reasonable to take toward doxastic practices we confidently engage in without being able to show that they are reliable.

So let's consider that question. And since the consideration will be much the same whichever doxastic practice we fasten on, we may as well stick with our old friend, SP.

A full consideration of this question is beyond the bounds of this essay, which is primarily designed to make the case, in a sufficiently detailed and thorough fashion, for the thesis that we do indeed have this problem with respect to SP, and to suggest that other basic sources of belief are faced with a similar difficulty. However, I can hardly leave the issue dangling in the air. I must at least lay out the alternatives open to us and briefly look into the reasons for preferring one to the others. That will be the task of the final chapter.

Chapter 5

WHERE DO WE GO FROM HERE?

i. The Problem

First, at the risk of belaboring the obvious, I will further underline the seriousness of the problem. As pointed out early in this essay, we are powerfully moved, both practically and theoretically, to seek to believe what is true. So strong are these motivations that we inevitably accord the value of true belief a preeminent place in our scale of values. It is extremely important to us to form beliefs in reliable ways, in ways that we can depend on to yield mostly true beliefs in the situations in which we typically find ourselves. Hence, when we reflect on our epistemic situation, we can hardly turn our backs on our inability to give a satisfactory demonstration of the reliability of SP and other doxastic practices in which we constantly engage. Complacency is not an option.

Nor is the classical skeptical stance of epoche, suspension of belief, a serious option. We are unable to refrain from forming beliefs by engaging in practices we are unable to show to be reliable without epistemic circularity. For, if the contentions of this essay are correct, that includes, directly or indirectly, all our common doxastic practices. And so, even more fundamental than the fact that we would not survive if we carried out so sweeping an abstention, there is the stubborn fact that to do so is not within our power. We lack any effective momentary control over belief formation. This is obvious with respect to the cases where something seems obviously true to us—whether on the basis of perception, introspection, memory, reasoning, or whatever. If it seems clear to me that I see a tree in front of me, I have no choice as to whether to believe that there is a tree in front of me. Even where the matter is not so clear, I do not think that one can believe one way or other, or withhold judgment, by an act of will. But however this latter may be, in most cases belief formation involves what seem to be obvious truths, and there our doxastic attitude is definitely not under momentary voluntary control.[1] To be sure, we can take measures to influence our beliefs. We can train ourselves to be more critical of gossip, we can selectively expose ourselves to considerations favoring one or another side of an issue, and so on. But it would be fantastic to suppose that such measures could lead to a total abstention from forming beliefs on the basis of perception, reasoning, etc. An across-the-board epoche is simply not a live option.

Some skeptically minded thinkers distinguish between 'belief' and what is variously called 'acceptance' or 'assent', and recommend suspension only of the latter.[2] "Assent" is construed as the conscious, deliberate acceptance of a proposition as true; and it is thought that we may withhold this, while still in a state of belief, in the sense that our behavior will still proceed as if the proposition in question were true. I doubt that "intellectual assent" can be divorced from behavioral dispositions to the extent that these thinkers suppose. I have no doubt but that many of our beliefs are held and acted on without the occasion for intellectual assent ever arising. The subject never asks herself the question that would call for such assent. But where I believe something, and this fact isn't hidden from myself in some way or other, I will assent to the proposition believed if the question is raised. If my belief is firm enough, I can't see that I have more of a choice here than with belief itself, though I can of course voluntarily go through the motions of saying to myself, "I don't know whether that is true or not". But waiving these doubts, let's agree for the sake of argument that intellectual assent is always under direct voluntary control. That is still not to the present point, which specifically concerns *belief*. Our present dilemma is that we are ineluctably engaged in forming beliefs in ways we cannot noncircularly show to be reliable. And that sticks in our craw, whatever we can do about "assent".

[1]See my "The Deontological Conception of Epistemic Justification" (Alston 1989b) for an extended discussion of the issue of voluntary control of belief.
[2]See, e.g., Moser 1989, pp. 15–16; Lehrer 1979, pp. 65–67.

If our situation requires us to make some active response to the epistemological crisis engendered by the results of this essay, and if abstention from the doxastic practices involved is not a possibility, what alternatives remain? I fear they are quite restricted. One possibility, already summarily dismissed in Chapter 1, is coherence theory. As pointed out there, a pure coherence theorist is not worried about the kind of circularity involved here. Within a total coherent system, the ultimate source of justification for its constituents, mutual support is the rule rather than the exception. It contributes to the coherence of the system if the totality of perceptual beliefs (and what is based on them) supports the thesis that perception is a reliable source of belief, *and* the latter supports the former. Since reciprocal support is the name of the game, it is no reproach to our confidence in the reliability of SP that it is needed to support what supports it. But, as also pointed out in Chapter 1, even if mutual support is an important aspect of respectable belief systems, it cannot be the whole story. If it were, we would be confronted with the specter of an indefinite plurality of equally coherent but mutually incompatible beliefs systems with no way of choosing between them. Coherence cannot be the sole, even the sole ultimate, contributor to positive epistemic status. The only escape from an unresolved plurality of incompatible belief systems is to suppose that some beliefs have a prima facie acceptability (justification, warrant . . .) independent of coherence considerations. That will give us a basis for preferring some coherent systems (the ones that have such beliefs) to others. And it is the conviction that, for example, SP confers such a prima facie acceptable status on its products that gives rise to our present difficulty.

Another possibility is to bite the bullet, embrace epistemic circularity and find some way of talking ourselves into regarding it as friend rather than foe, or at least as neutral. This, I take it, is the reaction of "naturalized epistemology", found in different forms in works by, for example, Willard Van Orman Quine (1969) and Alvin Goldman (1986).[3] On this approach, epistemology should not be construed as a "first philosophy"[4] that has the task of validating all our belief-forming procedures (including its own, part of what gets us

[3]For a variety of formulations, see Kornblith 1985.
[4]Quine 1969.

into the present bind). Rather, it should be thought of as enjoying basically the same status as the natural sciences, free to make use of anything we know (or justifiably believe) in addressing its questions. Thus, just as we can make use of sense perception, memory, reasoning, and so on, in chemistry, geology, or sociology without noncircularly establishing their reliability, so it is with epistemology. We can make use of what we have learned from SP, along with other things, in assessing the reliability of SP. We take our stand *within* SP and other doxastic practices that are firmly rooted in our lives in addressing whatever questions we happen to be dealing with. The investigation of knowledge, and of the justification or rationality of belief, no more depends on a justification of the reliability of its doxastic resources than does biology or engineering.

The obvious retort to this pronouncement is that epistemology is obviously different from the various natural sciences precisely in that part of its distinctive mission is to raise questions about the reliability of basic belief sources; and these questions, unlike the ones raised by physics or biology, cannot be shown to have a positive answer without running into epistemic circularity. That would seem to be a sufficient reason for being worried about an unjustified reliance on perceptual evidence in epistemology, while not being plagued with similar worries in the natural sciences. What is the naturalized epistemologist to say to this? He might attempt to retreat to the position considered earlier that whereas it is quite legitimate to question the reliability of, say, specific perceptual or memorial belief-forming mechanisms, or partial groups thereof, it is out of bounds to raise questions about the reliability of perception or memory in general. But what grounds could be given for this proscription other than the fact that if we try to do so we run into epistemic circularity? And that is just to repeat that we are faced with the problem with which we are currently trying to deal.

Again, he might hold that the reliability of our familiar basic doxastic practices is just a rock-bottom commitment from which there is no appeal. It is impossible to find anything more basic on the basis of which this commitment could be evaluated. I find this claim quite appealing, and it will play a major role in the response I shall shortly be advocating, though it is set there in a larger context that involves a kind of justification of it, as well as suggestions of how this commitment can be tested to a certain extent. But a totally uncritical acceptance of our customary practices, without any provision for rational rejection or modification, I find quite indefensible, provided, as I shall be arguing shortly, there is a possibility of rational criticism.

ii. A Practical Argument for the Rationality of SP

The alternative I will now present is, as I have just been hinting, closely related to the naturalized epistemology, bite-the-bullet approach. It is allied to the latter in taking it to be rational and proper to engage in our customary doxastic practices without having, or even being able to have, any positive noncircular reasons for supposing them to be reliable. In that respect, the positions agree in taking these familiar practices to be autonomous, acceptable on their own, just as such, without being adequately grounded on anything external. Since any attempt to show one of these practices to be reliable will, in effect, assume the reliability of some other of our familiar practices there is no appeal beyond those practices. We can't move a step without trusting the deliverances of one or more of these sources. Nevertheless, my "doxastic practice" approach to epistemology, as I call it, differs from the usual naturalized epistemology stance in two respects. First, it provides an argument of a sort, not for the reliability of SP, but for the rationality of engaging in SP and the rationality of taking SP to be reliable. Second, it recognizes that the intrinsic prima facie rationality of taking SP to be reliable can be overriden by various considerations, and can also be strengthened by what I call 'significant self-support'. Thus, even though every firmly established doxastic practice has a prima facie claim to be engaged in and to be regarded as rational, that claim is subject, in principle, to being disallowed by certain kinds of negative considerations and to be strengthened by others. We are not forced to accord rational approval to every established doxastic practice, no matter what. We are not necessarily totally locked into the status quo. I will now give a brief exposition of each of these points.

The argument for the rationality of engaging in, say, SP and of supposing it to be reliable is one that trades on a practical rationality.

The basic point is this. We have already seen that we cannot investigate the reliability of a given practice without engaging in that practice or some other(s) to obtain information we need for that investigation. And if we keep validating each practice by the use of others, we will find ourselves in a very small circle. Hence, looking at the whole picture, we will find ourselves relying on the practices under investigation for the facts adduced in support of the reliability of those practices. In the nature of the case, there is no appeal beyond the practices we find ourselves firmly committed to, psychologically and socially. We cannot look into any issue whatever without employing some way of forming and evaluating beliefs; that applies as much to issues concerning the reliability of doxastic practices as to any others. Hence there is no alternative to employing the practices we find to be firmly rooted in our lives, practices which we could abandon or replace only with extreme difficulty if at all. We have seen that the classical skeptical alternative of withholding belief altogether is not a serious possibility. In the press of life we are continually forming beliefs about the physical environment, other people, and how things are likely to turn out—whether we will or no. If we could adopt some basic way of forming beliefs about the physical environment other than SP, or some basic way of forming beliefs about the past other than memory (and it seems clear that this is not within our power), why should we? What possible rationale could there be for such a substitution? It is not as if we would be in a better position to provide a epistemically noncircular support for the reliability of these newcomers. The same factors that prevent us from establishing the reliability of SP, memory, and so on without epistemic circularity would operate with the same force in these other cases. Hence we are not in a position to get beyond, or behind, our familiar practices and definitively determine their reliability from a deeper or more objective position. Our human cognitive situation does not permit it.[5] Since we cannot take a step in intellectual endeavors without engaging in some

---

[5]Alvin Plantinga has suggested to me that this is not something distinctive of our *human* situation. Even God couldn't turn the trick. Even God couldn't *establish* the reliability of some belief-forming practice without using some doxastic practice to do so. And then the above regress argument would apply. However, granting that point, I, unlike Plantinga, think that God's situation in this regard is radically different from ours. Since, as I see it, God does not have *beliefs*, this particular problem could not come up for God. See my "Does God Have Beliefs?" (Alston 1989a).

doxastic practice(s) or other, what reasonable alternative is there to practicing the ones with which we are intimately familiar?

These considerations seem to me to indicate that it is eminently *reasonable* for us to form beliefs in the ways we standardly do. Even though we are debarred from giving a noncircular positive validation of these practices, we can indicate, as we have just done, that it is eminently reasonable for us to go along with our very strong, and perhaps even irresistible, inclination to form beliefs in these ways. At least these considerations yield an initial, prima facie rationality for continuance in such practices, pending sufficient overriding considerations, about which more in a moment.

But perhaps we have too hastily concluded that we have no reasonable alternative to forming beliefs in *all* the ways that are firmly entrenched in our lives. Why shouldn't we take our stand on one or more of these, and hold the others subject to judgment on that basis? This is what has often happened in the history of philosophy. Descartes takes his stand on what is rationally evident on reflection and takes sense perception to be under suspicion until it vindicates itself to rational intuition. Hume takes his stand on the awareness of "impressions and ideas" and relations amongst them, and takes sense perception of the external world and inductive reasoning to be questionable pending a verdict of innocent before the bar of impressions and ideas. To be sure, the results of this essay indicate that those who take this approach are going to be frustrated in their attempts to validate the secondary practices by exclusive reliance on the primary ones. But partisans of the Descartes-Hume approach will, no doubt, respond that they are proceeding rationally, and if their results are negative that just shows what reason dictates.

But are they truly following the dictates of reason? I would say that they are vulnerable to a charge of undue partiality in taking some of our firmly established doxastic practices for granted and requiring vindication of the others in the light of the former. Here are two expositions of this point by Thomas Reid (1970).

> The author of the "Treatise of Human Nature" appears to me to be but a half-skeptic. He hath not followed his principles so far as they lead him; but, after having, with unparalleled intrepidity and success, combated vulgar prejudices, when he had but one blow to strike, his courage fails him, he fairly lays down his arms, and yields himself a captive

to the most common of all vulgar prejudices—I mean the belief of the existence of his own impressions and ideas.

I beg, therefore, to have the honour of making an addition to the skeptical system, without which I conceive it cannot hang together. I affirm, that the belief of the existence of impressions and ideas, is as little supported by reason, as that of the existence of minds and bodies. No man ever did or could offer any reason for this belief. Descartes took it for granted, that he thought, and had sensations and ideas; so have all his followers done. Even the hero of skepticism hath yielded this point, I crave leave to say, weakly, and imprudently . . . what is there in impressions and ideas.so formidable, that this all-conquering philosophy, after triumphing over every other existence, should pay homage to them? Besides, the concession is dangerous: for belief is of such a nature, that, if you leave any root, it will spread; and you may more easily pull it up altogether, than say, Hitherto shalt thou go and no further: the existence of impressions and ideas I give up to thee; but see thou pretend to nothing more. A thorough and consistent skeptic will never, therefore, yield this point. To such a skeptic I have nothing to say; but of the semiskeptics, I should beg to know, why they believe the existence of their impressions and ideas. The true reason I take to be, because they cannot help it; and the same reason will lead them to believe many other things. (Pp. 81–82)

The skeptic asks me, Why do you believe the existence of the external object which you perceive? This belief, sir, is none of my manufacture; it came from the mint of Nature; it bears her image and superscription; and, if it is not right, the fault is not mine: I even took it upon trust, and without suspicion. Reason, says the skeptic, is the only judge of truth, and you ought to throw off every opinion and every belief that is not grounded on reason. Why, sir, should I believe the faculty of reason more than that of perception?—they came both out of the same shop, and were made by the same artist; and if he puts one piece of false ware into my hands, what should hinder him from putting another? (P. 207)

Reid's point could be put by saying that the only (noncircular) basis we have for trusting rational intuition and introspection is that they are firmly established doxastic practices, so firmly established that we "cannot help it"; and we have exactly the same basis for trusting sense perception, memory, nondeductive reasoning, and other sources of belief for which Descartes and Hume were demanding an external validation. They all "came out of the same shop", and therefore if one of them is suspect so are all the others.

A defender of Descartes' or Hume's procedure might well point to differences between, say, introspection and rational intuition on the one hand, and sense perception and nondeductive reasoning on the other. It will be claimed that whereas the latter practices not infrequently lead to incompatible results, this does not, or even, it is sometimes claimed, *cannot* happen with the former. It has been a familiar thesis that we are infallible with respect to our current states of consciousness, or at least that nothing could show that one has made a mistake about such matters.[6] And similar claims have been made for rational intuition. Reid himself recognizes important differences (1969, p. 296). However, I cannot see that such differences as exist warrant our accepting some of our firmly established doxastic practices without external support and not others. First, it is not at all clear that a judgment of infallibility or incorrigibility is justified for any of them. This has been a hotly debated topic with respect to introspection in recent decades,[7] and powerful arguments have been levelled against the ascription to it of incorrigibility, much less infallibility. For example, it seems *possible* that neurophysiology should progress to the point that we should have public evidence of a neurophysiological sort that a person's sensations, feelings, or thoughts, are different from what he sincerely believes them to be. As for rational intuition, one would be hard pressed to find a contemporary defender of its infallibility or incorrigibility, and for good reason. It can hardly lay claim to complete consistency of output. To some philosophers it has seemed self-evident that every event is causally determined; to others it has seemed self-evident that humans have free choice in a sense that is incompatible with the causal determinism of such choices. To some it has seemed self-evident that temporally backward causation is impossible; to others it has seemed self-evident that it is possible. And so it goes. If the deliverances of rational intuition contradict each other, it cannot be that they are all correct. Nor does it seem at all impossible, in some cases, to determine which party is correct. But if the strongest candidates for these "epistemic immunities" do not really enjoy them, then any epistemic superiority

[6]See, e.g., Hume 1888, I, iv, 2; Lewis 1946, pp. 182–83; Ayer 1959b, p. 59; Shoemaker, 1963, pp. 215–16. For other examples, see my "Varieties of Privileged Access" (Alston 1989a).

[7]For the negative position see, e.g., Armstrong 1968, chap. 6, sec. 10 and Aune 1967, chap. 2, sec. 1. Recently there has been a spate of psychological studies showing that introspection is not as surefire as has often been supposed.

that some of our basic practices enjoy over others is just a matter of degree and hardly warrants our taking some for granted and requiring others to justify themselves by the output of the privileged few.

Moreover, even if certain practices never, so far as we can tell, issue mutually contradictory outputs, while others do suffer this disability from time to time, this will still not justify the Descartes-Hume procedure of taking the former, but not the latter without external validation. For the fact remains that, however *internally* consistent the former may be, they still share with the latter the crucial feature of being insusceptible of a noncircular proof of reliability; and it is the latter with which we are centrally concerned. We have already pointed out, in our brief comments on coherence theory (p. 122), that internal consistency, or even a stronger kind of internal coherence, is radically insufficient to guarantee that the constituents of a belief system are by-and-large true. Hence, where *reliability* is in question, we lack sufficient excuse for treating different practices in a fundamentally different way.

Thus we will follow the lead of Thomas Reid in taking all our established doxastic practices to be acceptable as such, as innocent until proven guilty.[8] They all deserve to be regarded as prima facie rationally engaged in (or 'acceptable', as we shall say), pending a consideration of possible reasons for disqualification, reasons we shall go into shortly.

The reader may well have been struck by a similarity between this position regarding established doxastic practices and Wittgenstein's claims for "language games" in *On Certainty* (1969) that is even greater than the just noted similarity to Reid. What Wittgenstein calls 'language games' in *On Certainty* bear a marked resemblance to my doxastic practices; they are, or essentially involve, distinctive ways of forming beliefs. In more extended presentations of this position,[9] I spelled out the similarities with Wittgenstein. In both cases, there is an emphasis on the indefinite plurality of doxastic practices, their social establishment, their pre-reflective genesis, and their involvement in

---

[8]This amounts to a kind of negative coherentism for socially established practices; they do not require positive support in order to be (prima facie) acceptable, but only the absence of sufficient reasons against them. I am not at all tempted by a negative coherentism with respect to beliefs. But the considerations of this chapter provide powerful support, I believe, for such a position for doxastic *practices*.

[9]Alston 1989c; 1991a, chap. 4.

wider spheres of practice. And in both cases, there is the crucial insistence that, in a sense, there is no appeal beyond the practices we find ourselves engaged in. But that crucial point finds a much more uncompromising expression in Wittgenstein. He holds that there can be no way of subjecting established language games to rational criticism; and hence, on his verificationist principles, it is meaningless to ask whether such a practice is reliable, or whether its basic presuppositions are true, known, or justifiably believed. I reject both of these claims. Though, as should be abundantly clear by now, I insist that we cannot give an adequate noncircular demonstration of the reliability of basic doxastic practices, I do not take that to render them immune from all rational criticism, as I shall be explaining shortly. And even if I did, I wouldn't hold that we cannot intelligibly ask whether a given practice is reliable or whether its inherent presuppositions are true. I have no inclination to share Wittgenstein's verificationist predilections, with its attendant antirealist tendencies to hold that there is a distinct sense of terms like 'true' for each language game. If this is Wittgensteinianism, it is a nonverificationist, realist, un-Wittgensteinian form.

iii. Practical Rationality and Reliability

So far, I have been arguing that it is (practically) rational to engage in SP and other standard basic doxastic practices. But the original question was as to whether SP is reliable. Does the (practical) rationality of engaging in a practice have any bearing on whether it is reliable?

It may well be doubted that there is any such bearing. First, it is clear that the rationality of a practice does not *entail* its reliability. The claim to practical rationality was that where doxastic practices are firmly rooted in our lives, it would be folly to cease practicing them without very strong reasons for doing so; and we have no such reasons. This could be the case even if the practice were in fact unreliable. Moreover, the practical rationality of SP does not even provide nondeductive but sufficient grounds for supposing it to be reliable. I fail to discern any evidential tie; how could the practical rationality of engaging in SP be *evidence* for its reliability?[10] But then it looks as if the judgment that the practice is rational has no bearing on the likelihood that it will yield truths, in which case the argument for rationality will not advance our original aim of determining whether one or another practice is reliable.

Nevertheless, I believe that in showing it to be rational to engage in SP, I have thereby, not shown SP to be reliable, but shown it to be rational to suppose SP to be reliable. Let me explain. In what follows I will abbreviate 'it is rational to engage in SP' as 'SP is rational'.

In judging SP to be rational I am thereby committing myself to the rationality of judging SP to be reliable. Note the carefully qualified character of this claim. I did not say that in judging SP to be rational I was thereby *judging* it to be rational to suppose SP to be reliable, much less that I was thereby *judging* that SP is reliable. One can make the former judgment without making the latter. I may make the judgment of rationality without ever having raised the question of reliability and, hence, without having taken any stand on that issue. When I say that in judging that p I am thereby *committing* myself to its being the case that q, what I mean is this. It would be irrational (incoherent . . .) for me to judge (assert, believe) that p and deny that q, or even to abstain from judging that q *if the question arises*. The judgment that p puts me in such a position that either of those reactions would be irrational. There is no way in which I can both judge that p and take a doxastic attitude toward q other than acceptance.

It is in this sense that judging SP to be rational *commits* me to its being rational to suppose that SP is reliable. But how can this be? How can taking a stand on the practical rationality of SP put me in such a position? The reason is that we are dealing with *doxastic* practices, belief-forming practices. With many sorts of practices I can take it to be rational to engage in them without supposing them to enjoy the kind of success appropriate to them. I can take it to be rational to engage in playing squash for its health and recreational benefits, without thereby committing myself to the proposition that I will win most of my matches. But to engage in a doxastic practice is to form

---

[10]If it were, we would have to give up the central contention of this essay that there can be no adequate epistemically noncircular argument for the reliability of SP and other fundamental doxastic practices.

beliefs in a certain way. And to believe that p is to be committed to its being true that p.[11] But what is true of individual beliefs is also true of a general practice of belief formation. To engage in a certain doxastic practice and to accept the beliefs one thereby generates is to commit oneself to those beliefs being true (at least for the most part), and hence to commit oneself to the practice's being reliable. It is irrational to engage in SP, to form beliefs in the ways constitutive of that practice, and refrain from acknowledging them as true—and hence to refrain from acknowledging the practice as reliable—if the question arises.

But if one cannot engage in the practice and refuse to admit that the practice is reliable if the question arises, then in judging that the former is rational, one has committed oneself to the latter's being rational, in the sense of 'is committed to' we have been explaining. For I cannot hold that X is rational and coherently deny (or abstain from judging) that Y is rational, where accepting (engaging in) X commits me to accepting Y. If pursuing a Ph.D. commits me to the belief that it is possible for me to get a Ph.D., then I can't rationally hold both that it is rational to pursue a Ph.D. and that it is not rational to suppose that I can get a Ph.D. The rationality of a practice (action, belief, judgment . . .) extends to whatever that practice commits me to. But then if I judge SP to be rational and deny that it is rational to regard it as reliable, I would be in an incoherent situation. In judging SP to be rational, I am committed to judging it to be rational to suppose SP to be reliable.[12]

But then if I have shown, by my practical argument, that it is rational to engage in SP, I have thereby shown that it is rational to take SP to be reliable. For since the acknowledgement of the ra-

---

[11]Again this is 'is to be committed to' rather than 'is to believe that', for the very reason brought out in the preceding paragraph: one could believe that the sun is shining without having ever raised the question of truth, and even without having the concept of truth. But if one does raise the question, it is irrational to assert that p and abstain from asserting that it is true that p.

[12]Note that this is a case of what has been called 'pragmatic implication'. Indeed, pragmatic implication is felicitously characterized in just the terms we have used: it consists of the fact that *asserting* that p commits one to asserting that q, even though the *proposition* p, does not entail, or otherwise imply, the proposition q. Thus, in asserting that my car is in the garage I pragmatically imply that I believe that my car is in the garage, even though the propositions *my car is in the garage* and *I believe that my car is in the garage* are logically independent. This is just the situation we have with *SP is rational* and *it is rational to take SP to be reliable*.

tionality of the practice commits one to the rationality of supposing it to be reliable, to provide an adequate argument for the former will be to provide an adequate argument for the latter. Hence our argument from practical rationality, though it does not show that SP is reliable, does show that it is rational to take it to be reliable. No doubt, it would be much more satisfying to produce a direct demonstration of the truth of the proposition that *SP is reliable*. But since that is impossible, we should not sneer at a successful argument for the rationality of supposing SP to be reliable.[13]

It may be claimed that there is less than meets the eye in this conclusion. We have shown, at most, that engaging in SP enjoys a *practical* rationality; it is a reasonable thing to do, given our aims and our situation. But then it is only that same practical rationality that carries over, via the commitment relation, to the judgment that SP is reliable. We have not shown that it is rational in an *epistemic* sense that SP is reliable, where the latter involves showing that it is at least probably true that SP is reliable. This must be admitted. We have not shown the reliability attribution to be rational in a truth-conducive sense of rationality, one that itself is subject to a reliability constraint. But that does not imply that our argument is without epistemic significance. It all depends on what moves are open to us. If, as I have argued, we are unable to find noncircular indications of the truth of the reliability judgment, it is certainly relevant to show that it enjoys some other kind of rationality. It is, after all, not irrelevant to our basic aim at believing the true and abstaining from believing the false, that SP and other established doxastic practices constitute the most reasonable procedures to use, so far as we can judge, when trying to realize that aim.[14]

[13]This is analogous to the "fideist" move in religion. Pessimistic about the chances of directly establishing the truth of the existence of God, one seeks to show that it is rational for one to believe in God, as a postulate of pure practical reason, as a requirement for fullness of life, or whatever. But only *analogous* to fideism; I don't wish to wear that label.

[14]The point that it is a practical rather than an epistemic rationality that has been established (where it is a distinguishing mark of the latter that it implies likelihood of truth) enables us to meet an objection to our procedure from some recent attacks on the idea that a justification (or sufficient reasons, evidence, or grounds) for the belief that p is also a justification . . . for anything entailed by p. Dretske (1970) argues that my adequate perceptual evidence for 'That's a zebra' is not also adequate evidence for 'That's not a donkey with stripes painted on it', even though the former entails the latter. One might have parallel qualms about my claim that an adequate argument for p is also an adequate argument for everything to which the assertion of p commits one. If we were thinking of the rationality

iv. Overriders of Prima Facie Rationality

So much for the practical rationality argument for the rationality of engaging in such familiar doxastic practices as SP and for the rationality of taking them to be reliable. Finally, I must turn to a consideration of ways in which this rationality can be weakened or strengthened.

I have already made the point that the rationality which attaches to a doxastic practice just by virtue of its being firmly socially established is only a prima facie one, subject to being overriden by sufficient negative considerations. What sorts of negative considerations are both relevant and available? Let's continue to concentrate on reliability as far and away the most fundamental and important virtue of a belief-forming practice. That being the case, we can say that the most decisive way of nullifying the prima facie rationality of a socially established practice is to show it to be unreliable, or to present strong enough reasons for supposing it to be unreliable. But if the main contentions of this essay are correct, we are in no position to do this by a straightforward check on the accuracy of a proper sample of the outputs of the practice. For wherever epistemic circularity is unavoidable in arguing for the reliability of a practice (and these are the cases with which we are concerned here), we are unable to determine the truth value of outputs of the practice without relying, directly or indirectly, on outputs of that very practice. And so any attempt to directly establish unreliability by establishing the falsity of enough of the outputs would be self-defeating. We would be presupposing the reliability of the practice (in trusting its outputs) in order to establish its unreliability. Hence we must look for some more indirect way of piling up evidence for unreliability. Where are we to look for that?

The most obvious candidate is internal inconsistency. If two per-

---

involved in my discussion as guaranteeing likelihood of truth, this would be a serious objection. But since it is practical rationality that is involved, the situation is different. Dretske's point is that what renders probable the supposition that 'it's a zebra' does not also render probable the supposition that 'it's not a donkey with stripes painted on it'. But I am thinking of the rationality of SP as committing one not to the likelihood of the truth of the claim that SP is reliable (or anything entailing that), but rather the practical rationality of taking SP to be reliable, the thesis that, given what I have to go on, I am well advised to make and act on this assumption. Thus the qualms expressed by Dretske and Nozick (1981, chap. 3) do not apply here.

ceptual beliefs contradict each other, at least one is false. The existence of even one such pair is sufficient to show that SP is not perfectly reliable. And a large number of such pairs, relative to the total output, would show that SP is not sufficiently reliable for it to be rational to engage in it (if we have a choice in the matter). And it would certainly show that it is not rational to take SP to be reliable.[15] There is no doubt but that SP and other basic practices do generate mutually contradictory beliefs. Witnesses to crimes and automobile accidents often disagree as to what happened, and there are undoubtedly many other cases of disagreement that go unnoticed because they are of no practical importance. Switching to other basic doxastic practices, it is notorious that peoples' memories often conflict. And it is equally notorious that investigators not infrequently draw contradictory conclusions from the same data concerning such matters as the health hazards of a given pesticide. As noted above, it has often been claimed that introspection and rational intuition are in principle immune from any contradictions in their outputs, but the claim has as often been disputed.

Nevertheless, I doubt that any of our most basic doxastic practices yield enough mutually contradictory pairs to be disqualified as a rational way of forming beliefs. It is only on a fantastically rigoristic epistemology that one would be deemed irrational in holding perceptual beliefs just on the grounds that our ways of forming perceptual beliefs sometimes yield mutually contradictory beliefs. To be sure, what I am condemning as fantastic has often been held in the history of philosophy. Plato and Descartes, to name only two of the most prominent, have refused to allow that sense perception can be a source of knowledge on just that ground.[16] Nevertheless, along with

[15]At this point one might complain that it is *practical* rationality we have claimed to attach, prima facie, to any firmly socially established doxastic practice, while it is a more theoretical rationality that is to be denied to any practice that generates too many mutual contradictory pairs. But these modes of rationality are not walled off from each other by impermeable barriers. Though considerations other than truth can support practical rationality, as I have contended, truth considerations where available are also highly relevant to it. If we have solid evidence that SP is unreliable, then, given the overriding practical importance of having true rather than false beliefs, it is not even practically rational to engage in it, much less practically rational to suppose it to be reliable.

[16]Neither of these philosophers used anything very like our contemporary concept of justification. Perhaps if they had, they would not have denied justification to perceptual beliefs; it is hard to say.

most contemporary epistemologists, I take it to be the better part of wisdom to allow that sources of belief can be rationally tapped even if they sometimes yield contradictions, provided this is a small proportion of their output. However, the present point is that sufficiently extensive and persistent internal contradictions in the output of a practice would give us a conclusive case for regarding it as unreliable.

Furthermore, and for similar reasons, a massive and persistent inconsistency between the outputs of two practices is a good reason for regarding at least one of them as unreliable. It has been alleged that the whole of the output of SP comes into conflict with what has been established by rational intuition and deductive reasoning. Parmenides and Zeno contended that since SP represents things as multiple and as moving, it conflicts with what reason assures us about the nature of reality. F. H. Bradley argued that since SP represents its objects as interrelated spatiotemporally, none of its products are strictly true, since we can prove that any alleged relational complex is shot through with contradictions. I mention these historically famous cases not to endorse them but to illustrate claims to a massive interpractice inconsistency. Now, even if there are such inconsistencies, that does not tell us which of the conflicting parties is to be condemned. Rationalistic philosophers like those cited take it that SP is the loser, but that should not be taken for granted. To argue that it is self-evident that rational intuition should be trusted rather than SP when they conflict is obviously epistemically circular, and we certainly can't suppose that SP will support the choice of its rival. But where there is such conflict, we can infer that at least one of the contestants is unreliable.

What can we do to choose between the disputants in such a case? The only principle that suggests itself to me as both non-question begging and eminently plausible is the conservative principle that one should give preference to the more firmly established practice. What does being more firmly established amount to? I don't have a precise definition, but it involves such components as (a) being more widely accepted, (b) being more important in our lives, (c) having more of an innate basis, (d) being more difficult to abstain from, and (e) its principles seeming more obviously true. But mightn't it be the case in a particular conflict that the less firmly established practice is the more reliable? Of course that is conceivable. Nevertheless, in the absence of anything else to go on, it seems the part of wisdom to go with the more firmly established. It would be absurd to make the opposite choice; that would saddle us with all sorts of bizarre beliefs.[17]

What about interpractice conflicts that are less sweeping than those alleged by the likes of Parmenides and Bradley? Well, just as a modicum of intrapractice inconsistency is compatible with a degree of reliability sufficient for justification, so it is with interpractice contradiction. We do find quite a bit of this. Memories are not infrequently at variance with perceptible traces left by past events. "I would have sworn that I turned off the coffee pot, but I can't doubt the evidence of my senses that it is still on." Predictions arrived at inferentially are fairly often disconfirmed by direct perception; science would be much easier if this were not the case. But so long as these conflicts are no more frequent than is actually the case we can live with them. They do not entail such a degree of unreliability as to inhibit the justificatory force of both practices. Thus I would judge that none of the familiar practices we have been discussing are discredited by conflict with other practices.

Doxastic practices have fallen by the wayside, however, in the course of history and prehistory through being undermined by conflict with more firmly established rivals. This is what has happened with a great variety of magical practices and practices of divination. Predicting the future by scrutinizing the stars, tea leaves, or the entrails of sacred beasts, one forms beliefs that frequently come into conflict with what can be observed. The priest consults the entrails and predicts that enemy forces will appear before the city on the morrow, but the morrow comes and, as everyone can see, no enemy army is there.[18]

[17]To be sure, there is no advance guarantee that when two practices are in total or massive conflict with each other, one of them will be more firmly established than the other. But that is just an example of the general point that the human condition provides no guarantee that we will not be faced with questions we have no way to settle.

[18]In saying that such practices have fallen by the wayside, I am not claiming that no one engages in them or considers them to be reliable. The recrudesence of magic, witchcraft, and superstition of all kinds is a much advertised feature of our times. I am, rather, making a normative statement concerning their epistemic status. And even on the factual side it is true that such modes of belief formation are much less prevalent now than several thousand years ago, a result that is due in no small part to the frequent disconfirmation of their outputs.

v. Significant Self-Support

Next I want to consider a way in which the prima facie claims of established doxastic practices can be strengthened. We get the key to this by noting that not all epistemically circular arguments fall under the ban against track record arguments for being equally available for any doxastic practice. There are epistemically circular arguments that will help us to discriminate between practices since they cannot automatically be used for any practice whatever. To illustrate this, consider the following ways in which SP supports its own claims. (1) By engaging in SP and allied memory and inferential practices we are enabled to make predictions many of which turn out to be correct, and thereby we are able to anticipate and, to some considerable extent, control the course of events. (2) By relying on SP and associated practices we are able to establish facts about the operation of sense perception that show both that it is a reliable source of belief and why it is reliable. Our scientific account of perceptual processes shows how it is that sense experience serves as a sensitive indicator of certain kinds of facts about the environment of the perceiver.

These results are by no means trivial. It cannot be assumed that any practice whatever will yield comparable fruits. It is quite conceivable that we should not have attained this kind or degree of success at prediction and control by relying on the output of SP; and it is equally conceivable that this output should not have put us in a position to acquire enough understanding of the workings of perception to see why it can be relied on. To be sure, an argument from these fruits to the reliability of SP is still infected with epistemic circularity; apart from reliance on SP we have no way of knowing the outcome of our attempts at prediction and control, and no way of confirming our suppositions about the workings of perception. Nevertheless, this is not the trivial epistemically circular support that necessarily extends to every practice, the automatic confirmation of each output by itself. Many doxastic practices, like crystal-ball gazing, do not show anything analogous to the above results. Since SP supports itself in ways it conceivably might not, and in ways other practices do not, its claims to reliability are thereby strengthened; and if crystal-ball gazing lacks any comparable self-support, its claims suffer by comparison.

Analogous points can be made concerning memory, introspection, rational intuition, and various kinds of reasoning. The results we achieve by engaging in these practices and by using their fruits are best explained by supposing these practices to be reliable. One point to be made here is that the achievements we cited above as redounding to the credit of SP really also depend on the use of memory and reasoning of various sorts. Suppose we were restricted to what we could learn from individual perceptions. Suppose we were not able to store these bits of information in memory and retrieve them therefrom, generalize from this stored perceptual information, excogitate theoretical explanations and deductively derive observationally testable conclusions from them. In that case those impressive achievements of prediction and control, and that impressive insight into the mechanisms of perception would never be forthcoming. It is the reliance on all these doxastic practices that makes these achievements possible; and so their attainment provides significant self-support for all the doxastic practices that made an essential contribution. But in addition to that, other kinds of significant self-support are forthcoming for particular practices. For example, the combination of rational intuition and deduction yields impressive and stable abstract systems, some of which, particularly the mathematical ones, play a crucial role in scientific inquiry, in addition to their intrinsic value.

Since even significant self-support exhibits epistemic circularity, I will refrain from taking it to be an independent reason for supposing the doxastic practice in question to be reliable. Because self-support requires assuming the practice in question to be a reliable source of belief, it provides evidence for reliability only on the assumption of that reliability; and that is hardly evidence in any straightforward sense. Hence I am taking significant self-support to function as a way of strengthening the prima facie claim of a doxastic practice to a kind of practical rationality, rather than as something that confers probability on a claim to reliability. But as such it is by no means a negligible consideration.

Let me summarize the doxastic practice response I favor to the epistemic-circularity crisis. We can show (or argue effectively) that it is prima facie practically rational to engage in firmly socially established doxastic practices and to take them to be (reasonably) reliable. This status can be overriden by sufficient internal or external inconsistency in their outputs, and can be strengthened by significant self-support. It seems plausible to suppose that our most familiar and most basic belief-forming practices pass this test, and thus, absent serious inconsistency, that it is rational to form beliefs in accordance with them and to take them to be reliable sources of beliefs.

No doubt more can and should be said about this doxastic practice approach. As noted above, I had made a start at this (Alston 1989c and 1991a, chapter. 4), but a thoroughgoing development must await another occasion. This essay has been primarily devoted to a detailed defense of the thesis that we inevitably run into epistemic circularity when we seek to show that any of our basic ways of forming beliefs is reliable. At least I have given what I take to be a sufficiently detailed argument that this is the case with respect to what I have dubbed 'SP', the aggregate of our customary ways of forming perceptual beliefs about the environment. And I have suggested that an equally convincing argument for the same conclusion could be given for our other basic belief-forming practices. Given that situation, we are faced with an urgent question as to what attitude we should take toward ways of forming beliefs we can neither avoid nor show to be reliable. I have only suggested the lines along which an answer should be developed, a development that remains to be worked out.

BIBLIOGRAPHY

Alston, William P. 1982. "Religious Experience and Religious Belief". *Nous*, 16: 3–12.

———. 1989a. *Divine Nature and Human Language*. Ithaca, N.Y.: Cornell University Press.

———. 1989b. *Epistemic Justification*. Ithaca, N.Y.: Cornell University Press.

———. 1989c. "A 'Doxastic Practice' Approach to Epistemology". In *Knowledge and Skepticism*, ed. Marjorie Clay and Keith Lehrer. Boulder, Colo.: Westview Press.

———. 1991a. *Perceiving God*. Ithaca, N.Y.: Cornell University Press.

———. 1991b. "Higher Level Requirements for Epistemic Justification". In *The Opened Curtain* ed. Keith Lehrer and Ernest Sosa. Boulder, Colo.: Westview Press.

Armstrong, David M. 1968. *A Materialist Theory of the Mind*, London: Routledge & Kegan Paul.

———. 1973. *Belief, Truth, and Knowledge*. London: Cambridge University Press.

Aune, Bruce. 1967. *Knowledge, Mind, and Nature*. New York: Random House.

Ayer, A. J., ed. 1959a. *Logical Positivism*. New York: The Free Press.

———. 1959b. "Privacy". *The Proceedings of the British Academy*, 45: 43–65.

Bennett, Jonathan. 1966. *Kant's Analytic*. Cambridge: Cambridge University Press.

———. 1979. "Analytic Transcendental Arguments". In *Transcendental Arguments and Science*, ed. P. Bieri, R. P. Horstmann, and L. Krüger. Dordrecht: Reidel.

Bonjour, Laurence. 1985. *The Structure of Empirical Knowledge*. Cambridge, Mass.: Harvard University Press.

Bouwsma, O. K. 1949. "Descartes' Evil Genius". *Philosophical Review*, 58, pp. 141–51.

Brandt, Richard B. 1985. "The Concept of Rational Belief". *The Monist*, 68, no. 1.

Chisholm, Roderick M. 1977. *Theory of Knowledge*, 2d. ed. Englewood Cliffs, N.J.: Prentice-Hall.

———. 1989. *Theory of Knowledge*, 3d. ed. Englewood Cliffs, N.J.: Prentice-Hall.

Clarke, Samuel. 1738. *A Discourse Concerning the Being and Attributes of God, The*

Obligations of Natural Religion and the Truth and Certainty of the Christian Revelation. London: W. Botham.

Dennett, Daniel C. 1981. "True Believers". In *Scientific Explanation*, ed. A. Heath. New York: Oxford University Press.

Dore, Clement. 1984. *Theism.* Dordrecht: Reidel.

Dretske, Fred. 1970. "Epistemic Operators", *Journal of Philosophy*, 67: 1007–1023.

———. 1981. *Knowledge and the Flow of Information.* Cambridge, Mass.: MIT Press.

Feldman, Richard. 1985. "Reliability and Justification". *The Monist*, 68, no. 2: 159–74.

Foley, Richard. 1987. *The Theory of Epistemic Rationality.* Cambridge, Mass.: Harvard University Press.

Goldman, Alan. 1988. *Empirical Knowledge.* Berkeley and Los Angeles: U. of California Press.

Goldman, Alvin I. 1979. "What Is Justified Belief?". In *Justification and Knowledge,* ed. George S. Pappas. Dordrecht: Reidel.

———. 1986. *Epistemology and Cognition.* Cambridge, Mass.: Harvard University Press.

Grayling, A. C. 1985. *The Refutation of Skepticism.* Peru, Ill.: Open Court.

Hardin, C. L. 1988. *Color for Philosophers: Unweaving the Rainbow.* Indianapolis, Ind.: Hackett.

Hume, David. 1888. *A Treatise of Human Nature,* ed. L. A. Selby-Bigge. Oxford: Clarendon Press.

Kant, Immanuel. 1929. *Critique of Pure Reason,* trans. Norman Kemp Smith. New York: St. Martin's Press.

Kornblith, Hilary, ed. 1985. *Naturalizing Epistemology,* Cambridge, Mass.: MIT Press.

Lehrer, Keith. 1979. "The Gettier Problem and the Analysis of Knowledge". In *Justification and Knowledge,* ed. George S. Pappas. Dordrecht: Reidel.

Lewis, C. I. 1946. *An Analysis of Knowledge and Valuation.* La Salle, Ill.: Open Court.

Locke, John. 1975. *An Essay Concerning Human Understanding,* ed. P. Niddich. Oxford: Clarendon Press.

Malcolm, Norman. 1963. *Knowledge and Certainty.* Englewood Cliffs, N.J.: Prentice-Hall.

Mitchell, Basil. 1981. *The Justification of Religious Belief.* New York: Oxford University Press.

Moser, Paul. 1989. *Knowledge and Evidence.* New York: Cambridge University Press.

Nozick, Robert. 1981. *Philosophical Explanations.* Cambridge, Mass.: Harvard University Press.

Oldenquist, Andrew. 1971. "Wittgenstein on Phenomenalism, Skepticism, and Criteria". In *Essays on Wittgenstein,* ed. E. D. Klembe. Urbana: University of Illinois Press.

Plantinga, Alvin. 1974. *God, Freedom, and Evil.* Grand Rapids, Mich.: Eerdmans.

———. 1993. *Warrant and Proper Function.* New York: Oxford University Press.

Pollock, John. 1974. *Knowledge and Justification.* Princeton: Princeton University Press.

———. 1986. *Contemporary Theories of Knowledge.* Totowa, N.J.: Rowman & Littlefield.

Putnam, Hilary. 1981. *Reason, Truth, and History.* Cambridge: Cambridge University Press.

Quine, Willard van Orman. 1969. *Ontological Relativity and Other Essays.* New York: Columbia University Press.

Reid, Thomas. 1969. *Essays on the Intellectual Powers of Man.* Cambridge, Mass.: MIT Press.

———. 1970. *An Inquiry into the Human Mind.* Chicago: University of Chicago Press.

Rowe, William L. 1975. *The Cosmological Argument.* Princeton: Princeton University Press.

Shoemaker, Sydney. 1963. *Self-Knowledge and Self-Identity.* Ithaca, N.Y.: Cornell University Press.

Slote, Michael A. 1970. *Reason and Skepticism.* London: George Allen & Unwin.

Stich, Stephen. 1990. *The Fragmentation of Reason.* Cambridge, Mass.: MIT Press.

Strawson, P. F. 1959. *Individuals.* London: Methuen.

———. 1966. *The Bounds of Sense: An Essay on Kant's Critique of Pure Reason.* London: Methuen.

Swain, Marshall. 1981. *Reasons and Knowledge.* Ithaca, N.Y.: Cornell University Press.

Wittgenstein, Ludwig. 1953. *Philosophical Investigations,* trans. G. E. M. Anscombe. Oxford: Basil Blackwell.

INDEX

Library of Congress Cataloging-in-Publication Data
Alston, William P.
    The reliability of sense perception / William P. Alston.
        p.   cm.
    Includes bibliographical references and index.
    ISBN 0-8014-2862-9 (alk. paper). — ISBN 0-8014-8101-5 (pbk. :
  alk. paper)
    1. Senses and sensation.   2. Belief and doubt.   3. Knowledge,
  Theory of.   I. Title.
  BD214.A57   1993
  121'.3—dc20   92-54964